Innovation Meets Tradition in the Sheep and Goat Dairy Industry

Innovation Meets Tradition in the Sheep and Goat Dairy Industry

Editors

Pierluigi Caboni
Paola Scano

MDPI • Basel • Beijing • Wuhan • Barcelona • Belgrade • Manchester • Tokyo • Cluj • Tianjin

Editors
Pierluigi Caboni   Paola Scano
University of Cagliari   University of Cagliari
Italy   Italy

Editorial Office
MDPI
St. Alban-Anlage 66
4052 Basel, Switzerland

This is a reprint of articles from the Special Issue published online in the open access journal *Dairy* (ISSN 2624-862X) (available at: https://www.mdpi.com/journal/dairy/special_issues/Sheep_Goat_Dairy).

For citation purposes, cite each article independently as indicated on the article page online and as indicated below:

LastName, A.A.; LastName, B.B.; LastName, C.C. Article Title. *Journal Name* **Year**, *Volume Number*, Page Range.

ISBN 978-3-0365-4535-6 (Hbk)
ISBN 978-3-0365-4536-3 (PDF)

Cover image courtesy of Paola Scano

Contents

About the Editors

Pierluigi Caboni

Pierluigi Caboni, Ph.D., is full Professor at the University of Cagliari, Italy. He graduated from the University of Cagliari (Italy) in 1996 with a degree in Chemistry, he then obtained a PhD in Pathology and Environmental Toxicology (2001). His research project included the development of new analytical methods for the determination of botanical insecticides. From 2001 to 2005 Pierluigi worked at the University of California at Berkeley under the supervision of Prof. J. E. Casida with the research activity focusing on pesticide metabolism and pesticide molecular toxicology. From 2005 holds a position of Professor of Food Chemistry at the University of Cagliari in the High-Resolution Mass Spectrometry Laboratory, Department of Life and Environmental Sciences. His research activity is focused on metabolomics, plant-omics, food analysis, pesticide and drug metabolism.

Paola Scano

Paola Scano, Ph.D., is Professor of Physical Chemistry at the University of Cagliari, Italy. She graduated at the University of Cagliari (Italy) in 1986 with a degree in Biology. Supported by a 3-years grant of the NFCR (National Foundation Cancer Research), she worked at the University of St. Andrews, UK (1988–1990) and obtained a PhD in Theoretical Chemistry. Her research focusses on studying biological matrices, particularly food products, by applying spectroscopic methods and by analyzing big data with multivariate statistical analysis. She teaches spectroscopic methods for the study of food systems at the Faculty of Pharmacy at the University of Cagliari.

Editorial

Innovation Meets Tradition in the Sheep and Goat Dairy Industry

Paola Scano * and Pierluigi Caboni

Department of Life and Environmental Science, University of Cagliari, 09124 Cagliari, Italy; caboni@unica.it
* Correspondence: scano@unica.it

Citation: Scano, P.; Caboni, P. Innovation Meets Tradition in the Sheep and Goat Dairy Industry. *Dairy* **2021**, *2*, 422–424. https://doi.org/10.3390/dairy2030033

Received: 15 July 2021
Accepted: 3 August 2021
Published: 5 August 2021

Publisher's Note: MDPI stays neutral with regard to jurisdictional claims in published maps and institutional affiliations.

Small ruminants, such as sheep and goats, are mostly raised in smallholder farming systems widely distributed throughout the world. For some geographical areas, they cover an important economic, environmental and sociological role. Sheep and goats present some advantages over other large ruminants: their grazing preferences enable them to feed on weeds and shrubs; for their small size they require less space and are less likely to damage and compact soils; they are easier to work with and are cheaper to buy and maintain. The range of products produced by small ruminants is easy to market because demand is high, yet largely unfulfilled [1]. The ability to respond to the increasing market demands, in terms of the type of products or quality standards required, is crucial for the survival of the existing farming systems. Not only farmers, but the whole traditional ovine and caprine dairy product chain should be encouraged to adapt to meet consumer needs. In the last few decades, consumer demands have challenged toward higher productivity, sustainability, and safety, protecting at the same time the product's uniqueness. Innovation (implementing novel strategies in all the steps of the production chain) cannot prescind from the scientific research to test, control and validate novel strategies. Aimed at this goal, modern analytical platforms are blooming, often supported by sophisticated statistical data analysis. In this special issue, as a scientific contribution toward the innovation of the small ruminants dairy system, we addressed, by different approaches, issues such as effects of diet on milk quality, breeding systems, seasonality of milk production, uniformity of products, exportability, and shelf life.

Worldwide, sheep and goats are raised within a wide spectrum of feeding systems that lie within the two extremes: extensive vs. intensive. Extensive grazing refers to the use of large areas of unimproved natural land—rangeland—for free-roaming grazing livestock; on the contrary, in the intensive grazing, the animal feed comes mainly from artificial, seeded pastures. Today, in many areas, the traditional grazing on natural pastures of native flora systems has been ameliorated by pasture with selected plants, diet supplements, and other strategies for the animals' well-being, for environmental and economical sustainability and to overcome the seasonality of dairy supply. Given that the small ruminant husbandry is based mainly on smallholder farming, diet and hours spent outdoors by animals vary greatly without stringent protocols, making it difficult to assess and validate the best procedures. One of the studies of this Special Issue has tackled this issue, comparing the milk metabolite profiles of Sarda sheep bred in Sardinia (Italy) by two relatively small farmers under two grazing systems, that differed by the access to selected pasture, time of grazing and quantity of cereal grain (g/day/head) in the diet [2]. Multivariate statistical analysis of the GC-MS metabolite profile integrated with parameters obtained by the MIR calibration procedures, such as milk urea, allowed to highlight metabolites linked to the type of feeding system. Effects of the diet on milk from Sarda sheep bred in Sardinia were also addressed by the work of Manis et al. [3], where changes in milk components upon integration of diet with cocoa husks, an agricultural by-product of tropical countries, were evaluated. Agricultural by-products are fed to animals by over 80 percent of owners, the type and amount of which varies seasonally and depends on the region. In the Mediterranean basin, the effects of local agricultural by-products (grape, olive, tomato, citrus pulp and myrtle residues) in the diet of small ruminants is amply studied; however, today, the great availability of agro-industrial by-products produced worldwide opens the door to novel products to be tested. Manis et al. [3] assessed that diet integration with cocoa husks induced level changes in milk metabolites, as detected by UHPLC-QTOF-MS, implicated in the thyroid hormone metabolism and ubiquinol-10 biosynthesis. However, Manis et al. (2021) also observed that the major driver of metabolite changes was the sampling time; in fact, the design of experiment involved the collection of milk samples from treated and control groups along 4 weeks, thus spanning from the late spring to summer. Seasonality of milk composition is characterized by periods of "peak" milk production, after which milk yield declines in summer, as animals go from the mid to late stages of their lactation, and this is more marked in extensive breeding systems, due to changes in the quality and availability of natural pasture. Seasonality of milk fatty acid (FA) composition was studied by Nudda et al. [4], which turned their attention from the amply studied health beneficial omega-3 FA and conjugated linoleic acid (CLA) to the odd and branched chain classes of FA. The latter, long neglected, have recently sparked interest in the scientific community due to an inverse relationship with the development of human diseases. Besides the seasonality of the content of these FA in goat and sheep milk, their "transferability" to the derived cheese has been assessed. The detailed FA profile of cheese confirmed the higher nutritional quality of sheep cheese for beneficial FA, including odd and branched chain FA, compared to goat cheese, and the importance of the period of sampling in the definition of the FA profile.

In order to protect typicity of agricultural/food products, the European Commission has appointed different foods originating from a specific geographical area with the PDO (Protected Designation of Origin) label, which describes the product and reports a disciplinary to be followed. Heat treatment of milk is one of the specifics of cheese disciplinary; milk for cheese manufacturing can be raw or pasteurized/heat-treated. If not, specified manufacturers are free to apply their more convenient milk treatment. Fiore Sardo (PDO) cheese is the oldest ovine cheese of Sardinia (Italy), being historically produced by shepherds in very small artisanal cheese factories. It must be obtained exclusively using raw whole milk from the Sarda Sheep breed. However, the use of heat-treated milk for Fiore Sardo production has been associated with common industrial processing practices in spite of the specifications indicated in the PDO disciplinary. Among the analytical techniques able to discriminate cheese made from raw or heat-treated milk, Anedda et al. [5] have successfully tested the nuclear magnetic resonance (NMR) relaxometry. Water molecules in cheese could be described, at a first approximation, as either free or bound (hydration water molecules), and NMR can give an estimate of the two populations. Fiore Sardo from raw or heat-treated milk showed different percentages of the two water pools, with the latter cheese exhibiting more bounded water. One explanation is a shift of the calcium phosphate from soluble to colloidal state due to the thermal treatment. Micellar calcium phosphate is a highly hydrated colloid, and Ca and P association to casein and micelle hydration are strictly related phenomena that have a marked effect on NMR relaxometry. Going further, Anedda et al. [5] defined a new parameter that takes into consideration both the water population in the two states (free and bound) and the time it takes to relax (T2). This new parameter better discriminated raw from heated milk cheese and highlighted the low variability in industrial cheese samples. Among cheese, PDO Pecorino Romano, obtained from raw sheep milk, is one of the most popular and exported Italian ovine cheeses. The United States, the leading export destination, where this cheese is mainly used as an ingredient in the food industry, have doubts about the safety of Pecorino Romano produced from unpasteurized milk and the Food and Drug Administration (FDA) is frequently evaluating whether to propose more restrictive requirements on the sale of unpasteurized Pecorino Romano. Indeed, cheese made from unpasteurized milk can represent a risk for consumers due to the possible presence of some pathogenic bacteria. Lai et al. [6] monitored the survival of the pathogenic bacteria *Listeria monocytogenes*, *Salmonella* spp.,

Staphylococcus aureus, and *Escherichia coli* O157:H7 during ripening of Pecorino Romano cheese. Whole sheep milk was inoculated with bacteria and divided into two aliquots and only one underwent thermization, then Pecorino Romano was produced from the two batches of milk and the analyzed samples of cheese, collected after 1, 90, and 150 days of ripening, only samples differed for the milk heat treatment. After 24 h, a reduction in bacterial loads was observed for all pathogens in cheese produced from raw milk, while in cheese produced with thermized milk, the bacterial load was below detection. After 90 days of production, all the cheeses were microbiologically safe. Authors concluded that when Pecorino Romano cheese is produced under PDO specifications, either from raw or thermized milk, a combination of factors, including the speed and extent of curd acidification in the first phase of the production, together with an intense salting and a long ripening time, preclude the possibility of growth and survival of a number of pathogenic bacteria [6]. In the Mediterranean area, to revive the ovine dairy industry, the development of new fresh dairy products is increasing. In order to prolong their shelf life, contaminations from mold and yeast must be avoided. In the work of Scano et al. [7], the use of natural alternatives, such as autochthonous Lactobacillus strains, to synthetic preservatives has been tested on different species of mold. The strains were considered potential good candidates to be used in cheese manufacturing as bioprotective cultures. By a GC-MS metabolomics approach, authors highlighted those metabolites mostly involved in the antifungal activity in vitro. This research can be considered a further step towards the use of biopreservatives in the dairy industry.

For this valuable collection of research, editors would like to thank all the authors who submitted their papers to this special issue, the reviewers with their constructive comments, and the editorial staff of Dairy.

Conflicts of Interest: The authors declare no conflict of interest.

References

1. Pollott, G.; Wilson, R.T. Sheep and goats for diverse products and profits. *FAO Diversif. Bookl.* **2009**, *9*, 42.
2. Scano, P.; Carta, P.; Ibba, I.; Manis, C.; Caboni, P. An untargeted metabolomic comparison of milk composition from sheep kept under different grazing systems. *Dairy* **2020**, *1*, 30–41. [CrossRef]
3. Manis, C.; Scano, P.; Nudda, A.; Carta, S.; Pulina, G.; Caboni, P. LC-QTOF/MS Untargeted Metabolomics of Sheep Milk under Cocoa Husks Enriched Diet. *Dairy* **2021**, *2*, 112–121. [CrossRef]
4. Nudda, A.; Correddu, F.; Cesarani, A.; Pulina, G.; Battacone, G. Functional Odd-and Branched-Chain Fatty Acid in Sheep and Goat Milk and Cheeses. *Dairy* **2021**, *2*, 79–89. [CrossRef]
5. Anedda, R.; Melis, R.; Curti, E. Quality Control in Fiore Sardo PDO Cheese: Detection of Heat Treatment Application and Production Chain by MRI Relaxometry and Image Analysis. *Dairy* **2021**, *2*, 270–287. [CrossRef]
6. Lai, G.; Melillo, R.; Pes, M.; Addis, M.; Fadda, A.; Pirisi, A. Survival of Selected Pathogenic Bacteria during PDO Pecorino Romano Cheese Ripening. *Dairy* **2020**, *1*, 297–312. [CrossRef]
7. Scano, P.; Pisano, M.B.; Murgia, A.; Cosentino, S.; Caboni, P. GC-MS Metabolomics and Antifungal Characteristics of Autochthonous Lactobacillus Strains. *Dairy* **2021**, *2*, 326–335. [CrossRef]

 *dairy*

Article

An Untargeted Metabolomic Comparison of Milk Composition from Sheep Kept Under Different Grazing Systems

Paola Scano [1,*], Patrizia Carta [2], Ignazio Ibba [2], Cristina Manis [1] and Pierluigi Caboni [1]

[1] Department of Life and Environmental Science, University of Cagliari, via Ospedale 72, 09124 Cagliari, Italy; cristina.manis@outlook.it (C.M.); caboni@unica.it (P.C.)

[2] Regional Association of Sardinian farmers, Milk Analysis Laboratory, Loc. Palloni, Nuraxinieddu, 09170 Oristano, Italy; carta.patrizia@gmail.com (P.C.); ibba_ignazio@ara.sardegna.it (I.I.)

* Correspondence: scano@unica.it; Tel.: +39-347-344-1765

Received: 29 February 2020; Accepted: 2 April 2020; Published: 5 April 2020

Abstract: This study aimed to evaluate the effects of different feedings on main traits and polar and semi-polar metabolite profiles of ovine milk. The milk metabolome of two groups of Sarda sheep kept under different grazing systems were analyzed by gas chromatography coupled with mass spectrometry (GC-MS) and multivariate statistical analysis (MVA). The results of discriminant analysis indicated that the two groups showed a different metabolite profile, i.e., milk samples of sheep kept under Grazing System 1 (GS1) were richer in nucleosides, inositols, hippuric acid, and organic acids, while milk of sheep under Grazing System 2 (GS2) showed higher levels of phosphate. Statistical analysis of milk main traits indicates that fat content was significantly higher in GS1 samples while milk from GS2 sheep had more urea, *trans*-vaccenic acid, and rumenic acid. MVA studies of the associations between milk main traits and metabolite profile indicated that the latter reflects primarily the long chain fatty acid content, the somatic cell count (SCC), and lactose levels. All together, these results demonstrated that an integrated holistic approach could be applied to deepen knowledge about the effects of feeding on sheep's milk composition.

Keywords: GC-MS; metabolomics; feeding systems; sheep dietary supplement; ovine milk

---

1. Introduction

Compared to the rest of the world, where bovine milk represents the most common dairy source, the Mediterranean area is characterized by a significant number of sheep flocks designated to milk rather than meat production [1]. Sardinia, Italy, is one of the leading Mediterranean regions in the production of sheep dairy products. Sheep breeding and pecorino cheese manufacturing have strong economic and cultural links with this region. Though most of the herds in Sardinia are still under traditional grazing on natural pastures of native flora [1], pasture with selected plants and diet supplements have been introduced to ameliorate milk quality, the well-being of sheep, and for environmental and economical sustainability [2]. The effects of these feeding systems on milk quality should be studied. It is widely recognized that diet and season play a major role in modulating chemical composition of ruminant's milk. It has been assessed that the main components of sheep's milk appear to be less influenced by the type of farming system (indoors vs. outdoors) than by the feeding system (pasture alone vs. pasture plus feed supplements) [3], and milk urea measure has been successfully proposed to evaluate the energy contents of a diet [4]. Most of the studies on the effects of feeding systems on sheep milk quality are focused on the production of health beneficial fatty acids (FA), such as rumenic acid (C18:2 *cis*-9, *trans*-11) and linolenic acid (C18:3, *cis*-9,12,15) [5], as well as on the effects of polyphenols on milk quality and animal performances [2,6]. Differences in ovine milk and cheese composition due to barley and corn supplement to dry pasture have been recently studied by Caprioli et al. [7]. They assessed positive effects on milk retinol and alpha-tocopherol contents and cheese volatiles. Although several researches have studied different aspects of ovine milk main composition [1,8–11], there is still little information in literature regarding the metabolite profile of sheep's milk [12–14], as most of the works focused on cow's milk. One of the most valuable approaches for the investigation of metabolite profiles in a biological system is metabolomics [15]. Through the application of a gas chromatography coupled with mass spectrometry (GC-MS) based metabolomic approach, it has been possible to study different metabolic pathways linked to quality and other aspects of milk production [16–19], such as mastitis [12], diet [20], and ketosis [21]. In this study, by a GC-MS metabolomics approach, the milk metabolite profiles of sheep bred in Sardinia under different feeding systems were investigated in association with measured milk main traits such as fat, proteins, caseins, lactose, and urea.

2. Materials and Methods

2.1. Chemicals and Reagents

All the solvents were purchased from Sigma Aldrich, Darmstadt, Germany. Pyridine, methoxamine hydrochloride, potassium chloride, N-methyl-N-(trimethylsilyl) trifluoroacetamide (MSTFA), and analytical standards presented a purity greater than 95% (Sigma Aldrich, Darmstadt, Germany). Water was produced using a MilliQ purification system (Millipore, Milan, Italy).

2.2. Animals and Pasture

A total of seventy milk samples were collected from Sarda breed sheep. Animals belonged to two different flocks kept under two different grazing systems (GS1 and GS2) located in the same geographical area (coordinates 40°7′48.5″ N, 8°40′42.9″, E and 40°11′15.1″ N, 8°41′38.1″ E for GS1 and GS2, respectively) and underwent two different feeding systems (Table 1). In particular, one flock (GS2) was housed from 6 p.m. to 5 a.m. and during the day (10 a.m. to 14 p.m.), it had access to natural pasture while was allowed to graze in a selected pasture (*Avena sativa* 70%, *Hordeum vulgare* 15%, and *Trifolium incarnatum* 15%) only for 1–2 h in the afternoon. GS2 flock had ad libitum access to dry forage (from the selected pasture) during night. During the day, the GS1 flock had unlimited access to natural pasture. The natural pasture available for both flocks was composed by *Ferula communis, Thapsia garganica, Asphodelus ramosus, Pteridium aquilinum subsp. Aquilinum, Urginea maritima*, and *Trifolium repens*. Twice a day, both flocks received barley and corn grains (50/50 w/w) before milking (500 and 200 g/head/d for GS1 and GS2 groups, respectively), with an average dry matter intake (DMI) of 438 and 175 g/d/head for GS1 and GS2 groups, respectively (Table 1). The feeding systems were those usually adopted by the farmers, so they did not represent an additional management and manual labor or a stress for the animals. GS1 flock was composed of approximately 100 ewes and the GS2 flock of 120 ewes. The weight of animals was estimated as 43 ± 6 for GS1, and 40 ± 5 for GS2 and a body condition scoring (BCS) of 3.3 ± 0.2 for GS1 and 3.0 ± 0.2 for GS2.

Table 1. Description of the two diet systems.

Schedule	Grazing System 1 (GS1)	Grazing System 2 (GS2)
Day-grazing	Natural pasture [1] 10–12 h/d	Natural pasture [1] 4–6 h/d Selected pasture [2] 1–2 h/d
Milking	Barley and corn grains (50/50) 500 g/head/d DMI [3] = 437.5 g/head/d	Barley and corn grains (50/50) 200 g/head/d DMI = 175 g/head/d
Night		Forage *ad libitum*

[1] polyphite natural pasture: Ferula communis, Thapsia garganica, Asphodelus ramosus, Pteridium aquilinum subsp. Aquilinum, Urginea maritima, and Trifolium repens; [2] selected pasture for direct grazing and to obtain forage: 70% Avena sativa, 15% Hordeum vulgare and 15% Trifolium incarnatum; [3] DMI = dry matter intake.

2.3. Samples Collection

The morning individual milk samples were collected in spring 2019 from 70 pluriparous Sarda breed sheep in their late lactation period. Milk samples ($n = 37$ for GS1 and $n = 33$ for GS2) were collected in sterilized tubes through manual milking. Each milk sample was added with the preservative bronopol (2-bromo-2-nitro-1,3-propanediol) and divided in two aliquots. One was analyzed by the Sardinian Regional Animal Farmers Association (Associazione Regionale Allevatori, ARA, Sardegna) milk laboratory, and the other was stored at −20 °C for GC-MS analysis, which was performed within two days from collection. The milk samples were provided by the ARAS within the International Committee for Animal Recording (ICAR) program.

2.4. Milk Composition Analysis

Fat, proteins, caseins, lactose, and urea contents, freezing points (expressed in Horvet degree), and pH values were obtained with a Milkoscan FT6000 (Foss, Hilleroed, Denmark) following the procedure ISO 9622:2013 (ISO 9622:2013 IDF 141:2013). The fatty acid (FA) content was determined by Fourier-transform IR spectroscopy predictions as described by Caredda et al. [22]. Somatic cells were counted following the procedure ISO 13366-2-2006 (ISO 13366-2 IDF 148-2 2006) by a Fossomatic 5000 (Foss, Hilleroed, Denmark).

Milk yield was registered for each ewe at the morning milking and the total amount of milk produced by the two groups was recorded.

2.5. GC-MS Analysis

Thawed milk (1 mL) was subjected to ultrasounds for 15 min, and 0.1 mL were extracted following the Folch procedure [23] using 0.375 mL of a methanol and chloroform mixture (2/1, *v/v*). Samples were vortexed every 15 min up to 1 h, when 0.38 mL of chloroform and 0.09 mL of aqueous KCl 0.2 M solution were subsequently added. Samples were vortexed again and centrifuged for 15 min at 15294 g (Eppendorf 5810R, Milan, Italy). Two hundred µL of the hydrophilic supernatant were dried in glass vials using a nitrogen stream and then derivatized using 0.05 mL of methoxamine chloride dissolved in pyridine at 10 mg/mL, homogenized for 20 s, and kept at room temperature for 17 h. Then, 100 µL of MSTFA were added and samples were vortexed. After 1 h, 600 µL of hexane were added and samples homogenized again before GC-MS analysis. A Hewlett Packard 6850 gas chromatograph, a 5973 mass selective detector, and a 7683B series injector (Agilent Technologies, Palo Alto, CA, USA) were used to analyze samples using helium as the carrier gas at 1.0 mL/min flow. One µL of sample was injected in split-less mode and separated using a 30 m × 0.25 mm × 0.25 µm DB-5MS column (Agilent Technologies, Palo Alto, CA, USA). The temperatures for the inlet, interface, and ion source were 250 °C, 250 °C, and 230 °C, respectively. The oven temperature was programmed to increase from 50 to 230 °C by 5 °C/min over 36 min and kept at this temperature for 2 min. The mass range was set between 50 and 550 m/z using and electron voltage of 70. MSD ChemStation software (Agilent Technologies, Santa Clara, CA, USA) was used to elaborate the data. The GC-MS spectra deconvolution was performed by the AMDIS tool in the NIST08 library. Retention time and relative mass spectra of each compound were compared with the related analytical standards with the aim to identify those compounds previously recognized by consulting the NIST14 library of the National Institute of Standards and Technology (Gaithersburg, MD, USA), and the library developed at the Max Planck Institute of Golm (http://gmd.mpimp-golm.mpg.de/). Chromatograms in the AIA format were then uploaded to the XCMS Online platform [24]. The output of XCMS consisted in a 70 X 1230 matrix where each variable corresponded to the intensity value of an *m/z* ion at a specific retention time value. GC-MS features that annotated to bronopol with a retention time of 21.84 min were excluded.

2.6. Statistical Data Analysis

To obtain descriptive information on the data and to calculate the means and their standard deviations, we performed univariate analysis with the software OriginPro2016 (OriginLab, Northampton, MA, USA). When the two groups of samples were compared, the null hypothesis ("the means are not significantly different among the two sets of samples") was tested by using the unpaired samples two-tailed Student t-test. The output of XCMS pipeline was submitted to MVA, and analysis were performed as implemented in SIMCA-P+ software (version 14.1, Umetrics, Umeå, Sweden). Variables were mean centered and scaled to unit variance. At first, for sample distribution overview we performed a Principal Component Analysis (PCA) [25]. Subsequently, to identify discriminant metabolites between the two milk typologies, we performed an Orthogonal Partial Least-Squares-Discriminant Analysis (OPLS-DA). The obtained variable importance in projection (VIP) scores summarize the contribution of each variable to the model. In this work, GC-MS features showing VIP values greater than 1 underwent manually to a supervised analytes identification. A metabolite was considered significant only when at least two of its most abundant mass fragments and a retention time deviation <0.05 min were found in the list of VIP > 1. For quantification purposes for each metabolite, we considered the intensity of the most abundant mass fragment. Milk metabolite profiles were correlated to the measured main traits by a single-Y Partial Least-Squares (PLS) regression. The quality of the models was evaluated based on the cumulative parameters R^2Y and Q^2Y estimated by the default leave-1/7th-out cross validation in the corresponding PLS-DA models, as implemented in SIMCA-P+ program. The models were considered significant only when the difference between R^2Y and Q^2Y was <0.50 [26].

3. Results

3.1. Main Traits

The milk yield was higher in the sheep kept under grazing system 2 (GS2 flock, 784 ± 73 g/d) compared to (GS1 flock, 581 ± 64 g/d). The milk content of proteins, caseins, fat, lactose, FA, and other milk traits (expressed as mean ± SD) are reported in Table 2.

Milk of GS1 sheep showed a statistically higher fat content when compared to the GS2 group (5.9 ± 1.0 and 4.8 ± 1.2 g/100 mL, $p < 0.001$). An average milk fat level of 6-7 g/100 mL was reported for the Sarda breed [9]. It is well known that the milk fat content has a high variability, especially when compared with proteins and lactose levels, being strongly influenced by the feeding composition [3]. In the GS1 milk samples, consistently with their higher levels of fat, saturated and unsaturated long chain (C14-C18) FA linked to diet were found at higher levels, with the exception of n-3 FA. Contents of the de novo synthetized FA produced by the mammary gland (C4:0, C6:0, C8:0, C:10, C12:0) were found similar to those previously reported in literature for the Sarda breed sheep [27]. Differently from all other de novo FA, the concentration of C4:0 was statistically higher in GS1 samples. A different behavior of C4:0 with respect to the other de novo synthetized FA was already observed in sheep's [27] and cow's [28] milk. Genetic studies confirmed the independence of C4:0 from de novo mammary FA synthesis [29].

Among milk FA, due to their health benefits [30], a particular attention has been paid to rumenic acid, *trans*-vaccenic acid (C18:1 *trans*-11), and linolenic acid. Levels of *trans*-vaccenic acid (3.7 ± 0.5 *vs.* 3.0 ± 0.7 g/100 g of fat in the GS2 and GS1 groups, respectively) and rumenic acid (2.0 ± 0.2 *vs.* 1.6 ± 0.3 g/100 g of fat in the GS2 and GS1 groups, respectively) were significantly higher in GS2 milk. Milk products are the only food that naturally contain *trans*-vaccenic acid and rumenic acid produced by the ruminal biohydrogenation and the mammary Δ-9 desaturation pathways [1]. The pasture plays a pivotal role in increasing milk levels of *trans*-vaccenic acid and rumenic acid [5]. The preferred precursor of *trans*-vaccenic acid and of rumenic acid is the dietary linoleic acid. Oats, one of the plants in the selected pasture of GS2, is rich in linoleic acid and this fact can explain the higher content of *trans*-vaccenic acid and rumenic acid in the GS2 group.

Table 2. Descriptive statistics and means comparison of composition parameters of milk from sheep kept under grazing system 1 (GS1) and 2 (GS2).

	GS1 ($n = 37$)		GS2 ($n = 33$)		
Component	mean	SD [a]	mean	SD	p [b]
Fat (g/100 mL)	5.9	1.0	4.8	1.2	***
Proteins (g/100 mL)	5.6	0.6	5.8	0.4	
Caseins (g/100 mL)	4.3	0.5	4.4	0.4	
Lactose (g/100 mL)	4.8	0.4	4.9	0.3	
SCC [c] (10^3 cell/ mL)	1421		1337		
SCC (GM) [d]	119		175		
Urea (mg/100 mL)	24	5	41	8	***
Freezing point (milliHorvet)	−581	8	−582	7	
pH	6.9	0.1	6.8	0.1	
Chlorides (mg/100 mL)	141	46	129	32	
Total solids (g/100 mL)	16.4	1.5	15.5	1.2	**
SNF [e] (g/100 mL)	10.8	0.5	11.0	0.5	
Fatty acids (FA) (g/100 mL)					
Saturated FA	3.5	0.6	2.9	0.7	***
Unsaturated FA	1.8	0.3	1.4	0.4	***
Monounsaturated FA	1.4	0.3	1.1	0.3	***
Polyunsaturated FA	0.44	0.07	0.36	0.09	***
n-6 FA	0.12	0.04	0.09	0.04	***
n-3 FA	0.12	0.03	0.13	0.02	
C4:0	0.22	0.04	0.18	0.05	***
C6:0	0.15	0.03	0.15	0.03	
C8:0	0.12	0.03	0.12	0.03	
C10:0	0.32	0.09	0.35	0.10	
C12:0	0.16	0.05	0.17	0.04	
C14:0	0.5	0.1	0.4	0.1	***
C16:0	1.3	0.3	1.0	0.3	***
C18:0	0.6	0.2	0.4	0.2	***
C18:1	0.8	0.3	0.5	0.2	***
C18:2	0.13	0.04	0.11	0.03	***
C18:1 *trans*-11 (g/100g of fat)	3.0	0.7	3.7	0.5	***
C18:3 *n-3* (g/100g of fat)	1.7	0.3	1.7	0.2	
C18:2 *cis*-9, *trans*-11 (g/100g of fat)	1.6	0.3	2.0	0.2	***

[a] SD standard deviation; [b] p = probability student *t*-test, *** $p < 0.001$; ** $p < 0.01$; [c] SCC = somatic cell count; [d] GM = geometric mean; [e] SNF = solids-not-fat.

A parameter strictly connected with the quality of ewe's diet is the milk urea content, which is representative of urea levels of blood and other body fluids. Milk urea is a breakdown product of proteins and thus is considered a useful management tool to evaluate the metabolism and intake of proteins, minimizing potential negative effects on animals [4]. High milk urea content is associated with a worsening of animal welfare condition and different ruminant's pathologies, such as fertility and subclinical mastitis [31]. High urea content has been related both to the characteristics of the pasture with young sward poor in fiber and rich in highly fermentable proteins and/or non-protein nitrogen. High levels of dietary proteins can cause animal health and reproduction problems as well as the impairment of the milk characteristics and technological properties [4,32,33]. Sheep's milk of GS1 had significantly less urea than GS2 (24 ± 5 *vs.* 41 ± 8 mg/100 mL). However, both values are within the reported ovine milk urea level ranges [2,9], although some GS2 samples reached values of 60 mg/100 mL, which is a clear sign of unbalanced nutrition [2], while GS1 had values comparable with those found in summer season [9] when pasture is dry and poor. The higher urea content in GS2 can be due to a higher soluble protein content in the selected pasture and lower energy intake due to a probably insufficient grains ration.

3.2. Metabolomics

Forty-six polar and semi polar metabolites were identified and reported in Table S1. In Figure 1, the GC-MS chromatograms of milk samples from sheep kept under the two different grazing systems (GS1 and GS2) are shown. The former exhibits, in the range of 33–36 min, besides peaks annotated as lactose, a different chromatographic pattern, when compared to GS2, attributed to disaccharides such maltose and others. This fact is probably linked to a greater consumption of natural polyphite meadow, known to be rich in fiber, and to a higher portion of grains rich in starch in the GS1 feeding system (Table 1).

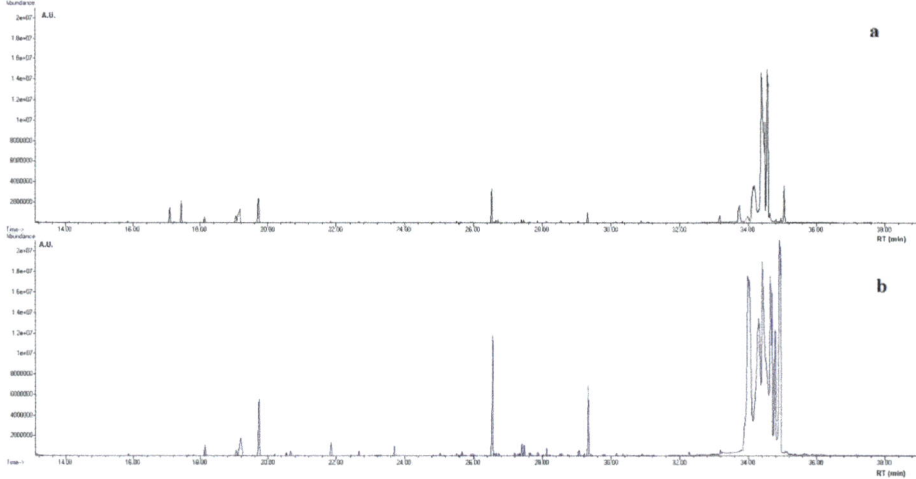

Figure 1. Representative GC-MS chromatograms of milk metabolites from sheep kept under grazing system 1 (GS1) (**b**) and grazing system 2 (GS2) (**a**).

To have an overview of samples similarities and dissimilarities we performed a PCA of GC-MS features. As shown in the score plot reported in Figure S1, milk samples clustered in two groups and GS2 milk samples were more scattered. Clustering indicated that the different feeding systems influenced the metabolite milk profile. In order to find discriminating metabolites between the two groups of sheep's milk samples, we performed a pair-wise OPLS-DA. The discriminant analysis correctly classified milk samples. Score plot is shown in Figure 2, and we report the metabolites that mostly contributed to sample classification in Table 3.

Further information can be obtained by associating the whole metabolite profile to the main traits, this approach can indicate whether modifications of a single trait are accompanied by modification of the whole metabolite profile. Results of the PLS analysis reported in Table 4 indicate that the metabolite profile is correlated to the content of the long chain FA, among which rumenic acid and *trans*-vaccenic acid, the most studied indexes of different pastures in the diet [5]. No *n-3* FA, neither de novo synthetized FA were found to be significantly linked to the metabolite profiles. Correlations between metabolites and SCC and lactose milk content were already observed [14,16].

Table 3. Pair-wise OPLS-DA[a] discriminant metabolites between milk from sheep kept under grazing system 1 (GS1) and grazing system 2 (GS2).

GS1		GS2	
Metabolite	VIP[b]	Metabolite	VIP
Maltose	1.86	Unk#1[c]	1.50
Myo-inositol	1.83	Unk#2	1.49
Malic acid	1.48	Phosphate	1.28
Hippuric acid	1.38		
Scyllo-inositol	1.27		
Succinic acid	1.25		
Uridine	1.24		
Glutaric acid	1.18		
Inosine	1.10		
Gluconic acid	1.09		

[a] Components = 1 + 2, R^2Y = 0.96, Q^2Y = 0.89; [b] VIP, variable importance in the projection of the OPLS-DA, only identified metabolites having VIP values > 1 are reported; [c] Unk, not annotated metabolite.

Table 4. Results of PLS correlating the whole GC-MS metabolite profile (n = 1200 features) to each composition main traits of sheep's milk samples (n = 70).

Component	R^2Y	Q^2Y	$R^2 - Q^2$
C18:2 *cis*-9, *trans*-11	0.95	0.64	0.31
C18:1	0.93	0.64	0.29
C18:1 *trans*-11	0.92	0.56	0.36
Monounsaturated FA	0.91	0.59	0.32
n-6 FA [a]	0.85	0.54	0.31
Polyunsaturated FA	0.85	0.54	0.31
C18:0	0.84	0.54	0.30
Unsaturated FA	0.84	0.56	0.28
SCC (log)	0.84	0.66	0.18
Lactose	0.83	0.61	0.22
Chlorides	0.81	0.59	0.22
pH	0.77	0.55	0.22
Urea	0.59	0.54	0.05
C18:2	0.35	0.24	0.11
C4:0	0.35	0.25	0.10
Fat	0.34	0.25	0.09
C16:0	0.33	0.25	0.08
Saturated FA	0.32	0.20	0.12
C18:3 *n-3*	0.30	0.14	0.16
C14:0	0.27	0.16	0.11
Total solids	0.23	0.12	0.11
Freezing point	n [b]		
SNF [c]	n		
n-3 FA	n		
C10:0	n		
C12:0	n		
Proteins	n		
Caseins	n		
C8:0	n		
C6:0	n		

[a] Fatty acids (FA); [b] n = analysis failed; [c] SNF = solids-not-fat.

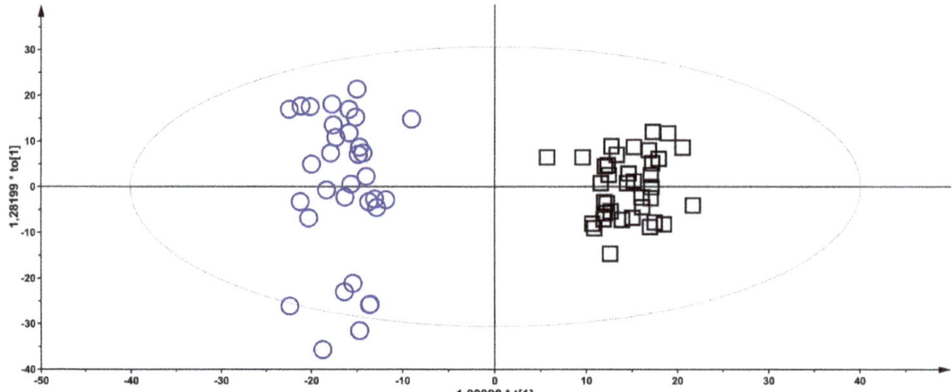

Figure 2. Pair-wise OPLS-DA score plot of GS2 ($n = 33$, empty circles) and GS1 ($n = 37$, empty boxes) milk samples, based on their GC-MS features. Components = 1 + 2, $R^2Y = 0.96$, $Q^2Y = 0.89$, discriminant metabolites along the predictive axis (*x*-axis) are reported in Table 3.

4. Discussion

GS1 milk samples showed higher levels of maltose, inosine, and uridine (Table 3). An increase of maltose and purine/pyrimidine cycle metabolites was observed in the rumen and milk of dairy cows fed with an increasing proportion of diet cereal grains [34–36], in agreement, as reported in Table 1, ewes in the GS1 received more concentrate than the GS2 group. Inosine and uridine belong to the non-protein-nitrogen (NPN) fraction of milk. These nucleosides are bioactive components with health beneficial effects [37]. Milk of sheep under GS1 showed higher levels of inositols (*myo*-inositol and *scyllo*-inositol, Table 3). Free inositols occur naturally in the environment as a bioactive component of living cells. Milk is a significant source of myo-inositol and for its role in neonatal nutrition, this polyol is often added to infant formulas to prevent a potential deficiency during early neonatal growth stage [38]. As well as urea, *myo*-inositol is osmolyte and the lower level of this polyol in GS2 can counterbalance their high urea content. Among the discriminant metabolites listed in Table 3, gluconic acid was found upregulated in the GS1 milk samples. Gluconic acid is the product of the glucose oxidation [39]. It has been demonstrated that in the rumen epithelium, butyrate (C4:0) and glucose oxidation mechanisms compete [40], and in the light of this observation we can hypothesize that the GS1 animals with a diet richer in concentrate and therefore in starch have more glucose available as substrate for the oxidizing enzymes, therefore saving more butyrate which is significantly higher in milk samples of the GS1 animals compared to GS2 (Table 2).

As shown in Table 3, hippuric acid, malic acid, and succinic acid were found in higher levels in GS1 milk samples. The content of hippuric acid in milk has been linked to the presence of polyphenols in forages [41]. Greater levels of hippuric acid in milk from pasture-based feeding systems have previously been reported in milk from goats and cows [19,36,42] and this molecule has been proposed as biomarker of pasture-derived milk [36,42]. In accordance, we found higher levels of hippuric acid in the milk of sheep allowed to graze on natural pasture for a longer time and thus used as biomarker for ovine milk. Malic acid and succinic acid are intermediates in the succinate and propionate pathway of ruminal bacteria for carbohydrate fermentation. Their higher levels in GS1 group could be linked to a greater consumption of natural polyphite meadow rich in fiber and grains when compared to GS2 group (see Table 1).

Compared to the GS1 samples reported in Table 3, milk of the GS2 group showed higher phosphorus content. A relationship between phosphorus and casein milk content is expected since this mineral is bound to casein micelles. Besides protein, lactose content can also affect the phosphorus content of milk since inorganic phosphorus is generated during the formation of lactose in the mammary gland and subsequently secreted into milk [43]. Although significant relationships between milk phosphorus and both protein and lactose contents were found [44], a considerable portion of the phosphorus content of milk remained unexplained, implying that other factors may affect milk phosphorus levels [45]. Feeding management is another important factor that can influence mineral concentration in milk. One source of organic phosphorus is phytate, which contributes to the majority of phosphorus in grains [46]. Ruminants can utilize phosphorus from phytate because ruminal microorganisms are capable of synthesizing phytase enzyme, which can release a phosphate group from the phytate [47]. Ruminal phytate hydrolysis is influenced by the type of grain, processing of feed ingredients, and supplemental exogenous phytase enzyme [48,49]. Lastly, as for the highly discriminant not-annotated metabolites (Unk#1 and Unk#2) characterizing GS2 milk samples, further studies will be carried out by GC-TOFMS to elucidate their chemical structure.

5. Conclusions

In summary, grazing system 1 led to a higher milk yield, more *trans*-vaccenic acid, and rumenic acid. Although accompanied by high urea levels, grazing system 2 was associated with higher fat and presence of metabolites linked to the grain supplements.

Our results suggest that the GC-MS metabolite profiles of ovine milk well reflect variations in diet with respect to pasture and concentrate supplementation and is strongly associate with the content of health beneficial FA. Further work with an ampler data set over a longer period of time will be carried out by this approach with the aim of establishing a model to classify milk and dairy products from sheep grazing natural pasture.

Supplementary Materials: The following are available online at http://www.mdpi.com/2624-862X/1/1/4/s1, Figure S1: PCA score plot, Table S1: GC-MS metabolite characteristics.

Author Contributions: Conceptualization, P.C. (Pierluigi Caboni), P.C. (Patrizia Carta), I.I. and P.S.; Methodology, P.C. (Pierluigi Caboni), I.I. and P.S.; Software, I.I. and P.S.; Validation, P.S., C.M. and I.I.; Formal Analysis, P.S., C.M. and I.I.; Investigation, PL.C, P.S. and I.I.; Data Curation, P.S., P.C., I.I.; Writing-Original Draft Preparation, P.C. (Pierluigi Caboni), P.S.; Writing-Review & Editing, P.C. (Pierluigi Caboni), P.S.; Visualization, P.S.; Supervision, P.C. (Pierluigi Caboni), P.C. (Paola Scano), I.I. and P.S.; Project Administration, P.C. (Pierluigi Caboni) All authors have read and agreed to the published version of the manuscript.

Funding: This research received no external funding.

Acknowledgments: We wish to thank the two sheep farmers Diego Deriu and Giovanni Deriu for collaborating with this research. This study was supported by the project: "Business Intelligent" Organic milk for the Chinese market. BIOmilkChina" POR FESR Sardegna 2014-2020. Asse I – Ricerca Scientifica, Sviluppo Tecnologico e Innovazione.

References

1. Scintu, M.F.; Piredda, G. Typicity and biodiversity of goat and sheep milk products. *Small Rumin. Res.* **2007**, *68*, 221–231. [CrossRef]
2. Molle, G.; Decandia, M.; Cabiddu, A.; Landau, S.Y.; Cannas, A. An update on the nutrition of dairy sheep grazing Mediterranean pastures. *Small Rumin. Res.* **2008**, *77*, 93–112. [CrossRef]
3. Morand-Fehr, P.; Fedele, V.; Decandia, M.; Le Frileux, Y. Influence of farming and feeding systems on composition and quality of goat and sheep milk. *Small Rumin. Res.* **2007**, *68*, 20–34. [CrossRef]
4. Cannas, A.; Pes, A.; Mancuso, R.; Vodret, B.; Nudda, A. Effect of dietary energy and protein concentration on the concentration of milk urea nitrogen in dairy ewes. *J. Dairy Sci.* **1998**, *81*, 499–508. [CrossRef]
5. Cabiddu, A.; Decandia, M.; Addis, M.; Piredda, G.; Pirisi, A.; Molle, G. Managing mediterranean pastures in order to enhance the level of beneficial fatty acids in sheep milk. *Small Rumin. Res.* **2005**, *59*, 169–180. [CrossRef]
6. Zervas, G.; Tsiplakou, E. The effect of feeding systems on the characteristics of products from small ruminants. *Small Rumin. Res.* **2011**, *101*, 140–149. [CrossRef]

7.	Caprioli, G.; Kamgang Nzekoue, F.; Fiorini, D.; Scocco, P.; Trabalza-Marinucci, M.; Acuti, G.; Tardella, F.M.; Sagratini, G.; Catorci, A. The effects of feeding supplementation on the nutritional quality of milk and cheese from sheep grazing on dry pasture. *Int. J. Food Sci. Nutr.* **2019**, *71*, 50–62. [CrossRef]

8.	Balthazar, C.F.; Pimentel, T.C.; Ferrão, L.L.; Almada, C.N.; Santillo, A.; Albenzio, M.; Mollakhalili, N.; Mortazavian, A.M.; Nascimento, J.S.; Silva, M.C.; et al. Sheep milk: Physicochemical characteristics and relevance for functional food development. *Compr. Rev. Food Sci. Food Saf.* **2017**, *16*, 247–262. [CrossRef]

9.	Scano, P.; Ibba, I.; Casula, M.; Contu, M.; Caboni, P. Effects of seasons on ovine milk composition. *J. Dairy Res. Tech.* **2019**, *2*, 004. [CrossRef]

10.	Park, Y.W.; Juárez, M.; Ramos, M.; Haenlein, G.F.W. Physico-chemical characteristics of goat and sheep milk. *Small Rumin. Res.* **2007**, *68*, 88–113. [CrossRef]

11.	Raynal-Ljutovac, K.; Lagriffoul, G.; Paccard, P.; Guillet, I.; Chilliard, Y. Composition of goat and sheep milk products: An update. *Small Rumin. Res.* **2008**, *79*, 57–72. [CrossRef]

12.	Caboni, P.; Manis, C.; Ibba, I.; Contu, M.; Coroneo, V.; Scano, P. Compositional profile of ovine milk with a high somatic cell count: A metabolomics approach. *Int. Dairy J.* **2017**, *69*, 33–39. [CrossRef]

13.	Caboni, P.; Maxia, D.; Scano, P.; Addis, M.; Dedola, A.; Pes, M.; Murgia, A.; Casula, M.; Profumo, A.; Pirisi, A. A gas chromatography-mass spectrometry untargeted metabolomics approach to discriminate Fiore Sardo cheese produced from raw or thermized ovine milk. *J. Dairy Sci.* **2019**, *102*, 5005–5018. [CrossRef] [PubMed]

14.	Caboni, P.; Murgia, A.; Porcu, A.; Manis, C.; Ibba, I.; Contu, M.; Scano, P. A metabolomics comparison between sheep's and goat's milk. *Food Res.* **2019**, *119*, 869–875. [CrossRef] [PubMed]

15.	Kim, S.; Kim, J.; Yun, E.J.; Kim, K.H. Food metabolomics: From farm to human. *Curr. Opin. Biotech.* **2016**, *37*, 16–23. [CrossRef]

16.	Goldansaz, S.A.; Guo, A.C.; Sajed, T.; Steele, M.A.; Plastow, G.S.; Wishart, D.S. Livestock metabolomics and the livestock metabolome: A systematic review. *PLoS ONE* **2017**, *12*, e0177675. [CrossRef]

17.	Melzer, N.; Wittenburg, D.; Hartwig, S.; Jakubowski, S.; Kesting, U.; Willmitzer, L.; Lisec, J.; Reinsch, N.; Repsilber, D. Investigating associations between milk metabolite profiles and milk traits of Holstein cows. *J. Dairy Sci.* **2013**, *96*, 1521–1534. [CrossRef]

18.	Melzer, N.; Wittenburg, D.; Repsilber, D. Integrating milk metabolite profile information for the prediction of traditional milk traits based on SNP information for Holstein cows. *PLoS ONE* **2013**, *8*, e70256. [CrossRef]

19.	Boudonck, K.J.; Mitchell, M.W.; Wulff, J.; Ryals, J.A. Characterization of the biochemical variability of bovine milk using metabolomics. *Metabolomics* **2009**, *5*, 375–386. [CrossRef]

20.	van Gastelen, S.; Antunes-Fernandes, E.C.; Hettinga, K.A.; Dijkstra, J. The relationship between milk metabolome and methane emission of Holstein Friesian dairy cows: Metabolic interpretation and prediction potential. *J. Dairy Sci.* **2018**, *101*, 2110–2126. [CrossRef]

21.	Klein, M.S.; Almstetter, M.F.; Nürnberger, N.; Sigl, G.; Gronwald, W.; Wiedemann, S.; Dettmer, K.; Oefner, P.J. Correlations between milk and plasma levels of amino and carboxylic acids in dairy cows. *J. Proteome Res.* **2013**, *12*, 5223–5232. [CrossRef] [PubMed]

22.	Caredda, M.; Addis, M.; Ibba, I.; Leardi, R.; Scintu, M.F.; Piredda, G.; Sanna, G. Prediction of fatty acid content in sheep milk by Mid-Infrared spectrometry with a selection of wavelengths by Genetic Algorithms. *LWT Food Sci. Technol.* **2016**, *65*, 503–510. [CrossRef]

23.	Folch, J.; Lees, M.; Sloane Stanley, G.H. A simple method for the isolation and purification of total lipids from animal tissues. *J. Biol. Chem.* **1957**, *226*, 497–509. [PubMed]

24.	Tautenhahn, R.; Patti, G.J.; Rinehart, D.; Siuzdak, G. XCMS Online: A web-based platform to process untargeted metabolomic data. *Anal. Chem.* **2012**, *84*, 5035–5039. [CrossRef]

25.	Cozzolino, D.; Power, A.; Chapman, J. Interpreting and reporting principal component analysis in food science analysis and beyond. *Food Anal. Methods* **2019**, *12*, 2469–2473. [CrossRef]

26.	Scano, P.; Murgia, A.; Pirisi, F.M.; Caboni, P. A gas chromatography-mass spectrometry-based metabolomic approach for the characterization of goat milk compared with cow milk. *J. Dairy Sci.* **2014**, *97*, 6057–6066. [CrossRef]

27.	Correddu, F.; Cellesi, M.; Serdino, J.; Manca, M.G.; Contu, M.; Dimauro, C.; Ibba, I.; Macciotta, N.P.P. Genetic parameters of milk fatty acid profile in sheep: Comparison between gas chromatographic measurements and Fourier-transform IR spectroscopy predictions. *Animal* **2019**, *13*, 469–476. [CrossRef]

28. Lock, A.L.; Preseault, C.L.; Rico, J.E.; DeLand, K.E.; Allen, M.S. Feeding a C16:0-enriched fat supplement increased the yield of milk fat and improved conversion of feed to milk. *J. Dairy Sci.* **2013**, *96*, 6650–6659. [CrossRef]

29. Pegolo, S.; Cecchinato, A.; Casellas, J.; Conte, G.; Mele, M.; Schiavon, S.; Bittante, G. Genetic and environmental relationships of detailed milk fatty acids profile determined by gas chromatography in Brown Swiss cows. *J. Dairy Sci.* **2016**, *99*, 1315–1330. [CrossRef]

30. Júnior, O.S.; Pedrao, M.R.; Dias, L.F.; Paula, L.N.; Coro, F.A.G.; De Souza, N.E. Fatty acid content of bovine milkfat from raw milk to yoghurt. *Am. J. Appl. Sci.* **2012**, *9*, 1300–1306.

31. Melendez, P.; Donovan, A.; Hernandez, J. Milk Urea Nitrogen and Infertility in Florida Holstein Cows1. *J. Dairy Sci.* **2000**, *83*, 459–463. [CrossRef]

32. Nudda, A.; Battacone, G.; Bencini, R.; Pulina, G. Nutrition and milk quality. In *Dairy Sheep Nutrition*; CABI Publishing: Wallingford, UK, 2004; pp. 129–149. ISBN 0-85199-681-7.

33. Pulina, G.; Nudda, A.; Battacone, G.; Cannas, A. Effects of nutrition on the contents of fat, protein, somatic cells, aromatic compounds, and undesirable substances in sheep milk. *Anim. Feed Sci. Technol.* **2006**, *131*, 255–291. [CrossRef]

34. Ametaj, B.N.; Zebeli, Q.; Saleem, F.; Psychogios, N.; Lewis, M.J.; Dunn, S.M.; Xia, J.; Wishart, D.S. Metabolomics reveals unhealthy alterations in rumen metabolism with increased proportion of cereal grain in the diet of dairy cows. *Metabolomics* **2010**, *6*, 583–594. [CrossRef]

35. Saleem, F.; Ametaj, B.N.; Bouatra, S.; Mandal, R.; Zebeli, Q.; Dunn, S.M.; Wishart, D.S. A metabolomics approach to uncover the effects of grain diets on rumen health in dairy cows. *J. Dairy Sci.* **2012**, *95*, 6606–6623. [CrossRef]

36. O'Callaghan, T.F.; Vázquez-Fresno, R.; Serra-Cayuela, A.; Dong, E.; Mandal, R.; Hennessy, D.; McAuliffe, S.; Dillon, P.; Wishart, D.S.; Stanton, C.; et al. Pasture feeding changes the bovine rumen and milk metabolome. *Metabolites* **2018**, *8*, 27. [CrossRef]

37. Schlimme, E.; Martin, D.; Meisel, H. Nucleosides and nucleotides: Natural bioactive substances in milk and colostrum. *Br. J Nutr.* **2000**, *84* (Suppl. 1), 59–68. [CrossRef]

38. Woollard, D.C.; Macfadzean, C.; Indyk, H.E.; McMahon, A.; Christiansen, S. Determination of myo-inositol in infant formulae and milk powders using capillary gas chromatography with flame ionisation detection. *Int. Dairy J.* **1996**, *37*, 74–81. [CrossRef]

39. Ramachandran, S.; Fontanille, P.; Pandey, A.; Larroche, C. Gluconic acid: Properties, applications and microbial production. *Food Technol. Biotech.* **2006**, *44*, 185–195.

40. Wiese, B.I.; Górka, P.; Mutsvangwa, T.; Okine, E.; Penner, G.B. Interrelationship between butyrate and glucose supply on butyrate and glucose oxidation by ruminal epithelial preparations. *J. Dairy Sci.* **2013**, *96*, 5914–5918. [CrossRef]

41. Besle, J.M.; Viala, D.; Martin, B.; Pradel, P.; Meunier, B.; Berdagué, J.L.; Fraisse, D.; Lamaison, J.L.; Coulon, J.B. Ultraviolet-absorbing compounds in milk are related to forage polyphenols. *J. Dairy Sci.* **2010**, *93*, 2846–2856. [CrossRef]

42. Carpio, A.; Bonilla-Valverde, D.; Arce, C.; Rodríguez-Estévez, V.; Sánchez-Rodríguez, M.; Arce, L.; Valcárcel, M. Evaluation of hippuric acid content in goat milk as a marker of feeding regimen. *J. Dairy Sci.* **2013**, *96*, 5426–5434. [CrossRef] [PubMed]

43. Shennan, D.B.; Peaker, M. Transport of milk constituents by the mammary gland. *Physiol. Rev.* **2000**, *80*, 925–951. [CrossRef] [PubMed]

44. Toffanin, V.; Penasa, M.; McParland, S.; Berry, D.P.; Cassandro, M.; De Marchi, M. Genetic parameters for milk mineral content and acidity predicted by mid-infrared spectroscopy in Holstein-Friesan cows. *Animal* **2015**, *9*, 775–780. [CrossRef] [PubMed]

45. Klop, G.; Ellis, J.L.; Blok, M.C.; Brandsma, G.G.; Bannink, A.; Dijkstra, J. Variation in phosphorus content of milk from dairy cattle as affected by differences in milk composition. *J. Agric. Sci.* **2014**, *152*, 860–869. [CrossRef]

46. Eeckhout, W.; De Paepe, M. Total phosphorus, phytate-phosphorus and phytase activity in plant feedstuffs. *Anim. Feed Sci. Tech.* **1994**, *47*, 19–29. [CrossRef]

47. Yanke, L.J.; Bae, H.D.; Selinger, L.B.; Cheng, K.J. Phytase activity of anaerobic ruminal bacteria. *Microbiology* **1998**, *144*, 1565–1573. [CrossRef]

48. Park, W.Y.; Matsui, T.; Yano, H. Post-ruminal phytate degradation in sheep. *Anim. Feed Sci. Tech.* **2002**, *101*, 55–60. [CrossRef]
49. Kincaid, R.L.; Garikipati, D.K.; Nennich, T.D.; Harrison, J.H. Effect of grain source and exogenous phytase on phosphorus digestibility in dairy cows. *J. Dairy Sci.* **2005**, *88*, 2893–2902. [CrossRef]

Article

Survival of Selected Pathogenic Bacteria during PDO Pecorino Romano Cheese Ripening

Giacomo Lai [1], Rita Melillo [2], Massimo Pes [1], Margherita Addis [1], Antonio Fadda [2] and Antonio Pirisi [1,*]

[1] Agris Sardegna, Servizio Ricerca Prodotti di Origine Animale, Bonassai, 07040 Olmedo, Italy; gialai@agrisricerca.it (G.L.); mpes@agrisricerca.it (M.P.); maddis@agrisricerca.it (M.A.)

[2] Istituto Zooprofilattico Sperimentale della Sardegna, Struttura Complessa Controllo Microbiologico e Ispezione degli Alimenti, Via Vienna 2, 07100 Sassari, Italy; rita.melillo@izs-sardegna.it (R.M.); antonio.fadda@izs-sardegna.it (A.F.)

* Correspondence: apirisi@agrisricerca.it; Tel.: +39-079-2842391

Received: 3 November 2020; Accepted: 2 December 2020; Published: 7 December 2020

Abstract: This study was conducted to assess, for the first time, the survival of the pathogenic bacteria *Listeria monocytogenes*, *Salmonella* spp., *Escherichia coli* O157:H7, and *Staphylococcus aureus* during the ripening of protected designation of origin (PDO) Pecorino Romano cheese. A total of twenty-four cheese-making trials (twelve from raw milk and twelve from thermized milk) were performed under the protocol specified by PDO requirements. Sheep cheese milk was first inoculated before processing with approximately 10^6 colony-forming unit (CFU) mL^{-1} of each considered pathogen and the experiment was repeated six times for each selected pathogen. Cheese composition and pathogens count were then evaluated in inoculated raw milk, thermized milk, and cheese after 1, 90, and 150 days of ripening. pH, moisture, water activity, and salt content of cheese were within the range of the commercial PDO Pecorino Romano cheese. All the cheeses made from raw and thermized milk were microbiologically safe after 90 days and 1 day from their production, respectively. In conclusion, when Pecorino Romano cheese is produced under PDO specifications, from raw or thermized milk, a combination of factors including the speed and extent of curd acidification in the first phase of the production, together with an intense salting and a long ripening time, preclude the possibility of growth and survival of *L. monocytogenes*, *Salmonella* spp., and *E. coli* O157:H7. Only *S. aureus* can be still detectable at such low levels that it does not pose a risk to consumers.

Keywords: cheese safety; foodborne pathogens; sheep milk; *Listeria monocytogenes*; *Salmonella* spp.; *Escherichia coli* O157:H7; *Staphylococcus aureus*; raw milk; thermization

---

1. Introduction

Consumer demand for unpasteurized milk cheeses is constantly increasing because of their more intense flavor and varied aroma than those of pasteurized milk cheeses [1–3]. However, especially when made from unpasteurized milk, cheese can hold a risk for the consumer, because of the possible presence of some pathogenic bacteria such as *Listeria monocytogenes*, *Salmonella* spp., *Staphylococcus aureus*, *Campylobacter* spp., *Brucella* spp., and pathogenic *Escherichia coli* [3–5]. Based on a data collection of dairy products-associated outbreaks in the United States from 1993 to 2006, Langer et al. [6] determined that outbreaks attributed to the consumption of unpasteurized dairy products were approximately 150 times more frequent, based on the unit of consumption than those related to pasteurized milk or pasteurized milk products. Likewise, Costard et al. [7] reported that in the United States, from 2009 to 2014, the consumption of unpasteurized dairy products caused 840 times more disease and 45 times more hospitalizations than that of pasteurized products.

Dairy 2020, 1, 297–312; doi:10.3390/dairy1030020   www.mdpi.com/journal/dairy

Among the pathogenic agents, *L. monocytogenes*, *Salmonella* spp., Shiga toxin-producing *E. coli* (STEC) as the serotype O157:H7 and enterotoxin-producing *S. aureus* are the most involved in foodborne outbreaks related to the consumption of raw milk cheese in industrialized countries [3,5]. These foodborne pathogens usually cause disease with acute symptoms restricted to the gastrointestinal tract such as diarrhea, abdominal cramps, nausea, vomiting, and limited in time and severity [8]. However, in some cases, they can cause serious diseases such as hemolytic uremic syndrome and thrombotic thrombocytopenic purpura associated to *E. coli* O157:H7 or meningitis and septicemia caused by *L. monocytogenes*, with a significant mortality rate in vulnerable groups such as infants, elderly, and immunocompromised adults [9,10].

These pathogenic bacteria can originate in raw milk by direct excretion from animals infected udder, by fecal contamination or from the farm environment more generally [5,11]. The foodborne pathogens can reach cheeses via contaminated milk or through the dairy plants environment and the processing equipment [4]. A recent survey shows how *L. monocytogenes* was first detected in a newly established cheese-making facility just nine months after the starting of the production [12]. Moreover, some pathogens like *L. monocytogenes* and *S. aureus* can colonize abiotic surfaces by forming biofilms, that make the bacteria immune to the action of antimicrobial agents [13,14]. Thus, they can persist for a long time in the manufacturing environment, where these microorganisms could be a potential cause of cross-contamination of the dairy products. Workers can also be an important source of contamination as a result of improper handling and some of them could be asymptomatic carriers of *S. aureus* [15,16].

The growth and survival of microbial pathogens during cheese-making is hindered by several factors such as milk heat treatments, curd cooking, rate of curd acidification by starter cultures, final product pH, salt addition, and competition with the native microflora present in milk [17]. Despite these hurdles, foodborne pathogens can express an adaptive response to different sublethal stresses, essential for their survival in harsher environments. Moreover, pathogenic bacteria adapted to a sublethal stress may exhibit cross-protection with enhanced resistance to different stresses [18–20]. Several challenge studies have shown that some foodborne pathogens are able to survive during the manufacturing and ripening of different kinds of cheese made from raw milk [9,21–23]. Some authors especially report that *L. monocytogenes*, *Salmonella* spp., and *E. coli* O157:H7 remained detectable after selective enrichment even for more than 200 days of ripening [24–26].

However, no studies have investigated the behavior of foodborne pathogens during the production and ripening of Pecorino Romano cheese. Pecorino Romano is an Italian protected designation of origin (PDO) semi-cooked hard cheese, which must be made exclusively from raw or thermized whole sheep milk, according to the PDO specifications [27]. PDO Pecorino Romano cheese is the most popular ovine cheese produced in Italy and it has a very important role from the economic point of view. Indeed, Pecorino Romano is one of the most exported Italian cheeses in the world [28]. The United States, with an export quota of around 13,000 tons in 2019, is the leading export destination, where this cheese is mainly used as an ingredient in the food industry [29].

Typically Pecorino Romano cheese has about 32% of moisture, a water activity (a_w) value of around 0.85, and a pH value between 5.07 and 5.31. This cheese usually has a high salt content that ranges from 4.5% to 8.3%. The minimum ripening period is 150 days for the table cheese, while grating cheese requires 240 days [28].

Although unpasteurized milk cheeses are commonly consumed in a large number of countries, some of them, the United States primarily, have doubts about the safety of these products. For instance, the Food and Drug Administration (FDA) is frequently evaluating whether to propose more restrictive requirements on the sale of unpasteurized milk cheeses, like Pecorino Romano cheese [30,31]. Despite the healthiness of this product, also indirectly attested from the lack of foodborne infection or intoxication episodes bound to the consumption of this cheese, experimental studies are important to investigate the fate of pathogenic bacteria during Pecorino Romano cheese manufacturing and ripening.

The main objective of this work was to investigate the survival of *L. monocytogenes*, *Salmonella* spp., *E. coli* O157:H7, and *S. aureus*, during the ripening of PDO Pecorino Romano cheese made from raw or thermized milk.

2. Materials and Methods

2.1. Experimental Design

Experimental productions of Pecorino Romano cheese were carried out according to PDO specifications, using cheese-making facilities of Agris Sardegna (Olmedo, Italy). Although PDO Pecorino Romano cheese is now exclusively produced from thermized milk, PDO specifications do not exclude the use of raw milk. Therefore, we performed two series of experimental cheese-makings, from raw milk (RM) and thermized milk (TM). Each series consisted of twelve cheese batches, three replicates for each selected pathogen. On the same day two experimental cheese-makings were performed starting from a single batch of raw whole sheep milk inoculated with a given pathogenic microorganism, the first one from raw milk and the second one after thermization of the same sheep milk.

2.2. Bacterial Strains and Inoculum Preparation

Seven strains of *L. monocytogenes*, two of *Salmonella* spp., three of *E. coli* O157:H7, and five of *S. aureus* were used (Table 1). We have chosen both reference and wild strains, the latter were isolated from milk and dairy products. Each inoculum was prepared from a culture containing different strains (reference and wild strains) of the same species.

Table 1. Pathogen strains used in Pecorino Romano cheese-making trials.

Microorganism	Strain [a]	Collection [b]
Listeria monocytogenes	ATCC 15313	ATCC [1]
	ATCC 19114	ATCC [1]
	ATCC 9525	ATCC [1]
	ATCC 153/3	ATCC [1]
	2	IZSLER [2]
	90	IZSLER [2]
	V7	IZSLER [2]
Staphylococcus aureus	ATCC 14458	ATCC [1]
	ATCC 25923	ATCC [1]
	401	IZS [2]
	466	IZS [2]
	64494	IZS [2]
Salmonella spp.	Typhimurium ATCC 6994	ATCC [1]
	Enteritidis 670	IZSLER [2]
Escherichia coli O157:H7	ATCC 43984	ATCC [1]
	47	IZLER [2]
	719	IZLER [2]

[a] Strain designation provided by collection. [b] Collection: ATCC, American Type Culture Collection, USA; IZSLER, Istituto Zooprofilattico Sperimentale della Lombardia e dell'Emilia Romagna, Italy; IZS, Istituto Zooprofilattico Sperimentale della Sardegna, Italy. [1] Reference strains. [2] Wild strains isolated from milk and dairy products.

Inoculum preparation included several steps. Briefly, all strains were conserved in Microbank beads (Biolife, Milan, Italy) at −18 °C. For each experiment, the strains were previously reactivated in brain heart infusion broth (Biolife, Milan, Italy). The pre-inoculation preparation included serial passages on trypticase soy broth (TSB, Oxoid, Thermo Fisher Scientific, Basingstoke, UK). Two milliliters of each strain culture were inoculated on 35 mL of TSB and incubated overnight at 37 °C under continuous stirring. Then, for each species (*L. monocytogenes*, *Salmonella* spp., *E. coli* O157:H7 and

S. aureus), subcultures have been combined in equal quantity and centrifuged at 6000 rpm for 10 min. Subsequently, the supernatant was discarded and the pellet was resuspended in physiological saline solution (NaCl 0.85%, pH 7). The optical density at 600 nm (Perkin Elmer Lambda 25 UV/VIS Spectrophotometer, Perkin Elmer, Waltham, MA, USA) was determined and counts were confirmed by serial decimal dilution and inoculation in the following selective agar media: Agar Listeria according to Ottaviani and Agosti (ALOA, Biolife, Milan, Italy) plates for *L. monocytogenes*; Baird Parker plates with Rabbit Plasma Fibrinogen supplement (RPF, Biolife, Milan, Italy) plates for *S. aureus*; cefixime tellurite sorbitol MacConkey agar plates (CT-SMAC, Biolife, Milan, Italy) for *E. coli* O157:H7; xylose lysine deoxycholate agar plate (XLD, Microbiol, Uta, Italy) for *Salmonella* spp. The agar plates were incubated at 37 °C for 24 h.

2.3. Cheese Production and Sampling

Each time, 700 L of raw whole bulk sheep milk collected from the "Bonassai" experimental farm of Agris Sardegna, were used. The milk was previously inoculated, under gentle stirring, with a multi-strain bacterial suspension of the same pathogenic microorganism to get a final concentration of approximately 10^6 CFU mL^{-1} and then was split into two 350 L aliquots, for RM and TM cheese-makings. In TM trials, milk was subsequently heated to 65 °C, without resting at the set temperature and quickly cooled down to coagulation temperature (38 °C). Then, the manufacturing process (RM and TM) followed the same procedure reported in Figure 1. Two vats were used alternately, one for RM trial and one for TM trial, in order to eliminate any vat variation. A thermophilic freeze-dried starter culture (FD-DVS CO-02, CHR Hansen, Hoersholm, Denmark) including *Streptococcus thermophilus* and *Lactobacillus delbrueckii* subsp. *bulgaricus* was added at a concentration of 10^6 CFU mL^{-1}.

Three cheese wheels of approximately 25 kg at 1 day were obtained from each cheese-making, for a total of 36 RM and 36 TM cheeses wheels. After molding, cheeses were subjected to drainage in hot room at 36 °C until reaching pH 5.20–5.30 and then at 20 °C up to 18–24 h. Dry salting and ripening were conducted in controlled conditions (10–12 °C and 78–85% relative humidity). In particular, cheeses were dry salted after 48 h (first application) from the manufacture. Later, during the first 90 days of ripening, cheeses were salted three more times, respectively after 12, 26, and 56 days. After 90 days, at the end of the salting, cheeses were washed, dried, and aged for two more months, for a total of 5 months of ripening. The cheese was sampled after 1, 90, and 150 days for physico-chemical analysis while inoculated raw milk, thermized milk, and cheese were sampled to enumerate pathogenic bacteria.

2.4. Physico-Chemical Analysis

All physico-chemical analysis were performed in duplicate. Samples of curd and cheese were analyzed for pH (pH-meter Basic 20+, Crison Instruments S.A., Alella, Spain). The following parameters were determined for the cheese samples: moisture (ISO 5534:2004) [32]; sodium chloride, determined by potentiometric titration with AgNO$_3$ (ISO 5943:2006) [33] (automatic titrator, model DL55, Mettler-Toledo GmbH, Schwerzenbach, Switzerland); water activity (a_w), determined at 25 °C (Aw Sprint instrument, Axair Ltd., Novasina Division, Lachen, Switzerland).

2.5. Microbiological Analysis

Aliquots of 25 mL or g for qualitative detection and 10 mL or g for quantitative detection were taken from each sample and used to prepare a suspension with appropriate diluents. The following parameters have been then researched:

(a) *Listeria monocytogenes.* Detection and enumeration were performed according to ISO 11290-1:2017 and ISO 11290-2:2017, respectively [34,35]. For detection, samples were mixed with Fraser broth base (Oxoid, Thermo Fisher Scientific, Basingstoke, UK), homogenized 90 s in a Stomacher Lab Blender 400 (International PBI S.p.A., Milan, Italy), and incubated for 24 h at 30 °C (primary enrichment). Subsequently, 100 µL of primary enrichment were transferred to

10 mL of Fraser broth supplemented by Fraser selective supplement (Oxoid, Thermo Fisher Scientific, Basingstoke, UK), which were incubated for 24 h at 37 °C (secondary enrichment). From primary and secondary enrichments, aliquots of 100 μL were streaked onto selective differential medium plates: Agar Listeria according to Ottaviani and Agosti (ALOA, Biolife, Milan, Italy) and Listeria selective agar (Oxford formulation, Oxoid, Thermo Fisher Scientific, Basingstoke, UK) and incubated for up to 24–48 h at 37 °C. All isolates with typical *L. monocytogenes* characteristics were subjected to morphological and biochemical proofs as confirmatory tests.

For the enumeration of *L. monocytogenes* ten-fold serial dilutions were made and aliquots of 100 μL were plated in duplicate on the surface of ALOA agar plates which were incubated for 48 h at 37 °C. Presumptive *L. monocytogenes* colonies were counted after confirmatory tests.

(b)   *Staphylococcus aureus.*   Detection and enumeration were performed according to ISO 6888-2:2004 [36]. Samples were weighed and mixed with buffered peptone water (BPW, Microbiol, Uta, Italy), ten-fold serial dilutions were prepared and aliquots of 100 μL were plated in duplicate on Baird Parker plates with Rabbit Plasma Fibrinogen supplement (RPF, Biolife, Milan, Italy) and incubated for 24–48 h at 37 °C.

(c)   *Escherichia coli* O157:H7. The detection was performed according to ISO 16654:2001/A1:2017 and four successive stages were necessitated [37].

   (1)   Enrichment of the test portion homogenized in modified tryptone soya broth containing novobiocin (mTSB + N, Biolife, Milan, Italy) with incubation at 41.5 °C for 6 h and subsequently for a further 12 h to 18 h.

   (2)   Separation and concentration of microorganisms by means of immunomagnetic particles coated with antibodies to *E. coli* O157:H7.

   (3)   Isolation by subculture of the immunomagnetic particles with adhering bacteria onto cefixime tellurite sorbitol MacConkey agar (CT-SMAC, Biolife, Milan, Italy) and sorbitol MacConkey agar (SMAC, Biolife, Milan, Italy) incubated at 37 °C for 24 h.

   (4)   Confirmation of typical colonies.

   E. coli O157:H7 count was determined by ten-fold serial dilutions and direct plating (100 μL in duplicate) on CT-SMAC agar plates incubated at 37 °C for 24 h. Typical *E. coli* O157:H7 colonies were subjected to confirmatory tests and were then enumerated.

(d)   *Salmonella* spp.   The detection was performed according to ISO method 6579-1:2017 [38]. The method required the following successive stages. A pre-enrichment in buffered peptone water (BPW, Microbiol, Uta, Italy) at 37 °C for 24 h. A selective enrichment in Rappaport-Vassiliadis with soy broth (RVS, Oxoid, Basingstoke, UK) and Müller-Kauffmann tetrathionate-novobiocin broth (MKTTn, Microbiol, Uta, Italy) for 24 h at 41.5 and 37 °C, respectively. Aliquots of the selective broths were streaked onto two selective isolation agar media, xylose lysine deoxycholate agar (XLD, Microbiol, Uta, Italy) and Salmonella detection and identification agar (SMID, BioMérieux, Marcy L'Etoile, France). The agar plates were incubated at 37 °C for 24 h. Confirmation of suspect colonies was carried out by biochemical and serological testing.

For the enumeration of *Salmonella* spp. ten-fold serial dilutions were prepared and aliquots of 100 μL were double plated on XLD agar plates which were incubated at 37 °C for 24 h. Presumptive *Salmonella* spp. colonies were subjected to confirmatory tests and were then counted.

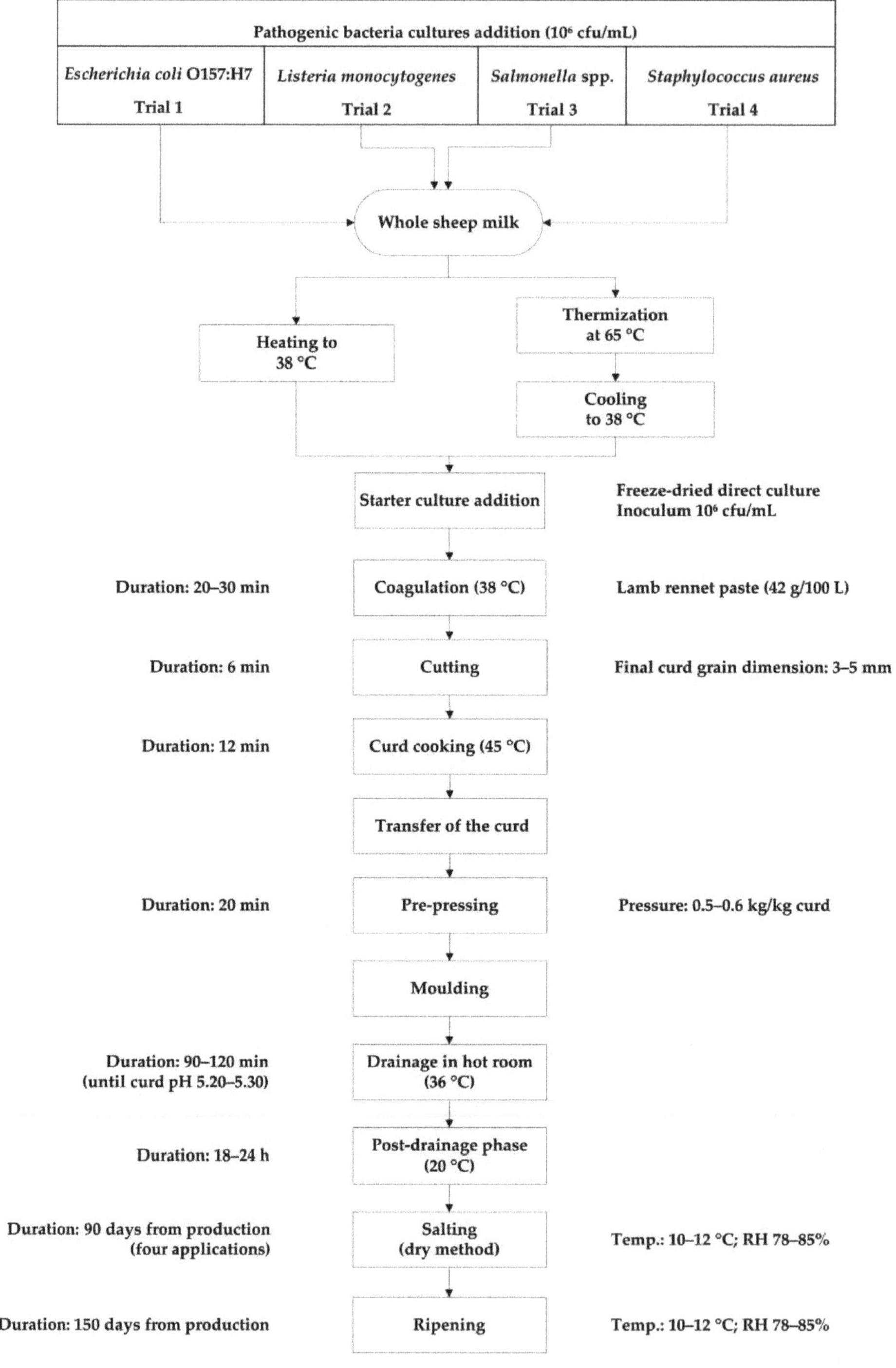

Figure 1. Experimental flow diagram of protected designation of origin (PDO) Pecorino Romano cheese production. RH = relative humidity.

2.6. Statistical Analysis

Microbial loads were log transformed and expressed as means and standard deviation (SD). The means and SD of physico-chemical parameters were determined from twelve cheese-makings for each experimental series (RM and TM). Analysis of variance was carried out using Minitab statistical package release 16 (Minitab Inc., State College, PA, USA). The general linear model procedure was used to verify the effects of the two studied factors "milk heat treatment" (2 levels) and "ripening time" (3 levels) on the microbial loads and physico-chemical parameters, as far as their interaction is concerned. The comparison between means was performed using Tukey's significant difference test ($p < 0.05$). Data were also analyzed by the Pearson correlation to measure the degree of the linear relationship.

3. Results and Discussion

3.1. Physico-Chemical Properties of Cheese

Figure 2 and Table 2 show the pH values in RM and TM cheese-making trials at different time. The pH values differed significantly between RM and TM ($p < 0.05$) in various step of cheese manufacturing process. As shown in Figure 2, after molding was completed, RM curd had a higher pH than TM curd (6.4 vs. 6.2, respectively; $p < 0.05$). This difference suggests a slight delay in the starter culture acidification activity in the early stage of RM cheese-making. This time lag could be due to the bacteriostatic activity of milk proteins such as lactoferrin, lactoperoxidase, and immunoglobulins since it normally decreases with milk heat treatments [39]. We cannot exclude that the competition of the starter culture with the native microflora of raw milk may also be involved in this delay of the acidification process. During drainage in hot room at 36 °C, a difference in pH values between RM and TM cheese-making was kept, however, this gap became less evident after two hours of drainage. Moreover, the lower SD in TM acidification curve shown in Figure 2 suggests a more regular and repeatable trend of the acidification process during TM cheese-making compared to RM one.

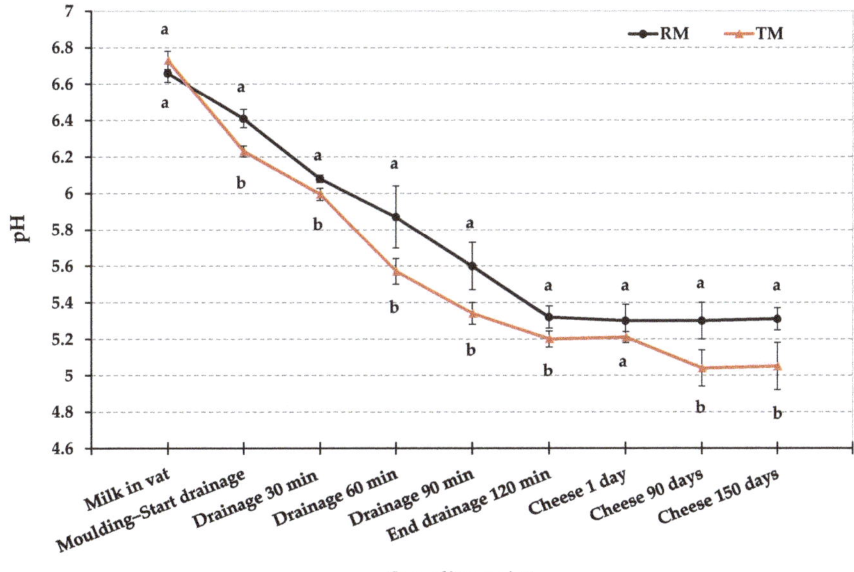

Figure 2. Acidification profiles of Pecorino Romano cheese produced from raw (RM), and thermized milk (TM). Twelve replicates for each cheese-making technology. Error bars indicate standard deviations. Different letters at the same time point indicate significant differences ($p < 0.05$).

Table 2. Physico-chemical parameters of Pecorino Romano cheese at different ripening times, obtained from raw (RM) and thermized milk (TM). Values are given as means ± standard deviation.

Ripening Time	Day 1		Day 90		Day 150		Significance		
Thermal Treatment	RM	TM	RM	TM	RM	TM	T	R	T × R
Parameters									
pH	5.3 ± 0.1 [a]	5.21 ± 0.03 [a]	5.3 ± 0.1 [a]	5.0 ± 0.1 [b]	5.3 ± 0.1 [a]	5.0 ± 0.1 [b]	**	**	**
Moisture (%)	42 ± 1 [a]	42 ± 1 [a]	32 ± 1 [bc]	32.5 ± 0.3 [b]	32 ± 2 [bc]	31.2 ± 0.4 [c]	NS	***	*
A_w	0.970 ± 0.005 [b]	0.983 ± 0.002 [a]	0.89 ± 0.01 [c]	0.880 ± 0.005 [cd]	0.880 ± 0.02 [cd]	0.873 ± 0.004 [d]	NS	***	***
NaCl/DM (%)	0.15 ± 0.01 [c]	0.21 ± 0.01 [c]	6.3 ± 0.2 [b]	7.2 ± 0.3 [a]	6.4 ± 0.2 [b]	7.3 ± 0.2 [a]	***	***	***

Values in the same line with different superscript letters differ significantly (Tukey's test, $p < 0.05$). DM = dry matter; A_w = water activity; RM = cheese from raw milk; TM = cheese from thermized milk; T = thermal treatment; R = ripening time. NS, $p > 0.05$; * $p < 0.05$; ** $p < 0.01$; *** $p < 0.001$.

As shown in Table 2, the pH values in RM and TM cheeses were significantly affected by the milk thermization treatment ($p < 0.01$) and ripening time ($p < 0.01$), furthermore the interaction between these factors was significant ($p < 0.01$). Despite the slight delay in the acidification process, after 1 day, RM and TM cheeses were not statistically different for pH values (5.3 and 5.21, respectively). On the contrary, after 90 and 150 days of ripening, RM and TM cheeses differed significantly for pH values ($p < 0.01$). In particular, pH in RM cheeses remained substantially unchanged during ripening (around 5.3), while pH in TM cheeses showed a slight decrease, reaching an average pH value of 5.0 at 150 days. The evolution of pH during cheese ripening is due to several factors involved in the metabolism of the main components of the cheese: hydrolysis of the residual lactose, which involves a decrease in pH, and degradation of proteins that, instead, leads to an increase in pH value [40,41]. Pecorino Romano cheese is characterized by modest proteolysis due to the intense salting that limits the activity of the proteolytic enzymes. On the other hands, lipolysis is more accentuated mainly because of the use of exogenous lipases contained in the lamb paste rennet, used for milk coagulation. These enzymes catalyze the biochemical process of triglyceride hydrolysis resulting in the release of short-chain fatty acids that can contribute to a decrease in the pH during ripening [28,42]. Finally, it is important to point out that the pH values found both for RM and TM cheeses are in the range of variation found in commercial PDO Pecorino Romano cheese [28].

In Table 2 we report also the remaining physico-chemical properties of RM and TM cheese, at different time of ripening. The milk thermization treatment did not significantly affect the moisture content, which instead was significantly influenced by ripening time and its interaction with heat treatment (respectively, $p < 0.001$ and $p < 0.05$). No significant differences were found between RM and TM cheese at each observed ripening time. The major changes in moisture occur in the first 90 days of ripening in both TM and RM cheeses, while the decrease from 90 up to 150 days was statistically significant only in TM cheese ($p < 0.05$). Moisture content was approximately 42% in cheese after 1 day and 31–32% in cheese after 90 and 150 days of ripening, according to Addis et al. [28,42].

As reported for moisture content, also the a_w values were significantly affected by the ripening time and its interaction with heat treatment ($p < 0.001$) (Table 2). The a_w, significantly differed in RM and TM 1 day cheeses, and dropped significantly from approximately 0.98–0.97 (TM and RM, respectively) up to 0.89–0.88 at the end of the salting process (90 days), then slightly decreased tendentially ($p > 0.05$) at 150 days. The a_w values at the end of ripening (150 days) were higher than those reported by Addis et al. [28]. However, these authors referred to longer aged Pecorino Romano cheese (7–8 months).

At the end of the salting process (90 days), the NaCl on dry matter (DM) content was significantly lower in RM cheeses (6.3%) compared to TM cheeses (7.2%). Salt content slightly increased, up to the end of ripening (6.4% and 7.3%, respectively in RM and TM cheeses). This difference in NaCl content between RM and TM cheeses could be due to the intrinsic variability of the dry salting technique. However, NaCl values are comparable to those of commercial Pecorino Romano cheese, which is characterized by high and variable salt content [28].

These results showed that the experimental cheeses met the typical requirements of PDO Pecorino Romano cheese. Moreover, the different cheese-making technology (RM and TM) seems to affect only in part the characteristics of the product at the end of the ripening period, which substantially has the same peculiarities of the Pecorino Romano cheese available on the market.

3.2. Pathogenic Bacteria Counts in RM Cheese-Making Trials

The pathogens counts in raw milk are given in Table 3. Pathogenic bacteria levels in raw milk pointed out the accuracy of the technique used for the inoculum preparation. Indeed, all the counts were around 6 $\log_{10}$ CFU mL^{-1}, as established from the experimental protocol. Table 4 shows the pathogens counts in cheese at different ripening stages. A reduction in bacterial loads was observed for all pathogens in 1-day old cheese. *L. monocytogenes* and *Salmonella* spp. showed the major reduction, with a decrease of about 4 $\log_{10}$. The number of viable *E. coli* O157:H7 decreased by about 2.5 $\log_{10}$, while *S. aureus* counts showed a reduction by about 0.5 $\log_{10}$. The behavior of the pathogenic bacteria was likely in part attributed to the number of viable cells lost with the whey separation and partly to the combined effect of various hurdles occurring during the first phase of Pecorino Romano cheese manufacturing, such as competition with the starter culture, acidification, and curd cooking [17,43]. Conversely, other studies showed an increase in pathogens counts during the early stages of cheese-making in different unpasteurized milk cheeses. Bellio et al. [21] reported a growth of 2 log cycle in *E. coli* O157:H7 levels during the first 24 h of PDO Fontina cheese production, despite a curd cooking phase at 48 °C. Similar findings were also given by other authors in Cacioricotta, Cheddar, and Gouda cheeses [9,24]. Chatelard-Chauvin et al. [26] found that *L. monocytogenes* increased by about 3.5 $\log_{10}$ in Cantal cheese at 24 h. An increase in *Salmonella* spp. populations over 1 $\log_{10}$ was observed in Gouda and Cheddar cheeses in the first 24 h of manufacturing [25,44]. In our study, *S. aureus* fell slightly, however, some authors observed a growth in *S. aureus* counts in the early stages of cheese production [45,46]. These authors considered that the increase in pathogenic bacteria counts during the early stages of cheese-making could be due both to an actual bacterial growth and to the entrapment of pathogens cells within the curd, during curd contraction and whey separation. In our study, the reduction in pathogens population levels during the early stages of cheese manufacturing might be related not only to the aforementioned hurdles to bacteria growth and survival but also to a protracted curd breaking (about 6 min) into extremely fine grains (3–5 mm). This is a typical feature of Pecorino Romano cheese-making technology which could lead to a higher level of pathogens viable cells lost in the cheese whey.

All tested pathogens, despite a reduction, were still present in 1-day old cheese in relevant bacterial loads. However, it must be pointed out that in the present study, high inoculum levels were used, which are difficult to find in practice, to simulate the worst-case scenario of contamination. Overall, our results are in agreement with other studies that indicate the ability of the pathogenic bacteria to overcome the obstacles of the first phase of cheese-making, first of all, acidification, probably owing to an acid tolerance response (ATR) widely reported in the literature [18–20].

Table 3. Results of pathogens counts ($\log_{10}$ CFU mL^{-1}) in sheep milk before and after thermization. Values are given as means ± standard deviation. *Escherichia coli* O157:H7: ATCC 43984, 47, 719. *Salmonella spp.*: Typhimurium ATCC 6994, Enteritidis 670. *Listeria monocytogenes*: ATCC 15313, ATCC 19114, ATCC 9525, ATCC 153/3, 2, 90, V7. *Staphylococcus aureus*: ATCC 14458, ATCC 25923, 401, 466, 64494.

Pathogens	Raw Milk	Thermized Milk	Significance
E. coli O157:H7	6.1 ± 0.1 [a]	2.6 ± 0.5 [b]	***
Salmonella spp.	6.5 ± 0.4 [a]	3.3 ± 0.4 [b]	***
L. monocytogenes	6.01 ± 0.02 [a]	3.6 ± 0.3 [b]	***
S. aureus	6.5 ± 0.1 [a]	3.7 ± 0.5 [b]	***

Values in the same line with different superscript letters differ significantly (Tukey's test, $p < 0.05$). *** $p < 0.001$.

Table 4. Results of pathogens counts ($\log_{10}$ CFU g^{-1}) in Pecorino Romano cheese at different ripening times, obtained from raw (RM), and thermized milk (TM). Values are given as means ± standard deviation. *Escherichia coli* O157:H7: ATCC 43984, 47, 719. *Salmonella spp.*: Typhimurium ATCC 6994, Enteritidis 670. *Listeria monocytogenes*: ATCC 15313, ATCC 19114, ATCC 9525, ATCC 153/3, 2, 90, V7. *Staphylococcus aureus*: ATCC 14458, ATCC 25923, 401, 466, 64494.

Ripening Time	Day 1		Day 90		Day 150		Significance		
Thermal Treatment	**RM**	**TM**	**RM**	**TM**	**RM**	**TM**	**T**	**R**	**T × R**
Pathogens									
E. coli O157:H7	3.5 ± 0.1 [a]	0 [b]	0 [b]	0 [b]	0 [b]	0 [b]	***	***	***
Salmonella spp.	2.3 ± 0.3 [a]	0 [b]	0 [b]	0 [b]	0 [b]	0 [b]	***	***	***
L. monocytogenes	2.01 ± 0.02 [a]	0 [b]	0 [b]	0 [b]	0 [b]	0 [b]	***	***	***
S. aureus	5.9 ± 0.3 [a]	<1 [b]	<1 [b]	<1 [b]	<1 [b]	<1 [b]	***	***	***

Values with different superscript letters differ significantly (Tukey's test, $p < 0.05$). RM = cheese from raw milk; TM = cheese from thermized milk; T = thermal treatment; R = ripening time. *** $p < 0.001$.

Table 4 also shows the pathogens counts at the end of salting (90 days) and the end of the ripening period (150 days). At the experimental conditions, all inoculated pathogens except *S. aureus*, were not detectable after 90 days, even after selective enrichment. This result was confirmed in 150-days old cheese. *S. aureus* counts were below the detection limit for direct plating enumeration method (<1 $\log_{10}$ CFU g^{-1}), both in cheese at 90 and 150 days of ripening. In the early phase of cheese-making (24 h), the pathogenic bacteria were subjected to initial stress because of the curd cooking and fast acidification. Afterwards, in the first 90 days of ripening, a significant correlation was found between microbial loads and a_w, moisture and NaCl content. In particular, all the studied pathogens exhibited a strong positive correlation with moisture (E. coli O157:H7, $R^2 = 0.987$, $p < 0.001$; L. monocytogenes, $R^2 = 0.982$, $p < 0.001$; *Salmonella* spp., $R^2 = 0.987$, $p < 0.001$; S. aureus, $R^2 = 0.993$, $p < 0.001$) and a_w (*E. coli* O157:H7, $R^2 = 0.971$, $p < 0.001$; L. monocytogenes, $R^2 = 0.982$, $p < 0.001$; *Salmonella* spp., $R^2 = 0.971$, $p < 0.001$; S. aureus, $R^2 = 0.948$, $p < 0.001$). Conversely, a strong negative correlation was found between microbial count and salt content (*E. coli* O157:H7, $R^2 = -0.998$, $p < 0.001$; L. monocytogenes, $R^2 = -0.997$, $p < 0.001$; *Salmonella* spp., $R^2 = -0.987$, $p < 0.001$; S. aureus, $R^2 = -0.996$, $p < 0.001$).

Several studies show the ability of pathogenic bacteria to survive in cheeses even beyond 90 days of ripening. Nevertheless, none of these had similar physico-chemical properties to those of PDO Pecorino Romano cheese (pH ≤ 5.4; moisture ~32%; a_w < 0.90; NaCl 4–8%), together with a long ripening period. Ioanna et al. [9] reported that *E. coli* O157:H7, when inoculated in milk at 2 $\log_{10}$ CFU mL^{-1}, survived in 90 days-old Cacioricotta cheese with a bacterial load greater than 4 $\log_{10}$ CFU g^{-1}. Similar findings were observed in Fontina cheese at 80 days of ripening [21]. In Cheddar and Gouda cheeses, *E. coli* O157:H7 was detectable after an enrichment procedure for up to around 300 days, with an initial raw milk inoculation of about 20 CFU mL^{-1}, which was far lower than that used in the present study [24]. *E. coli* O157:H7 appears to be one of the pathogens having the greatest ability to overcome environmental stresses and it is also infectious at very low doses (5–50 viable cells) [47]. Therefore, proving its complete inactivation is relevant. Viable cells of *Salmonella* spp. were detectable in Gouda cheese for more than 200 days [25]. However, in Feta and Tulum cheese, with a salt content close to that of Pecorino Romano cheese, *Salmonella* spp. was no more found after 20 and 90 days, respectively [48,49]. *L. monocytogenes* was detectable for more than 250 days in Cantal cheese [26]. In agreement with our study, Wusimanjiang et al. [50] observed that *L. monocytogenes*, when inoculated in milk at 3 $\log_{10}$ CFU mL^{-1}, reached undetectable levels after 28–35 days of ripening in a cheese with a salt content similar to that of Pecorino Romano cheese (4–10% on DM basis).

According to Regulation (EC) No 2073/2005, Pecorino Romano cheese, already at 90 days, belongs to ready-to-eat foods that do not support the growth of *L. monocytogenes*, since its a_w value is less than 0.92. For these products, the Regulation establishes a tolerance for *L. monocytogenes* up to 100 CFU g^{-1} [51]. The cheeses we obtained were *Listeria*-free at 90 days, despite the high initial inoculum in cheese milk.

S. aureus, unlike the other pathogens, was still present in high population levels in 1-day old cheese. Since *S. aureus* is halotolerant, it could find the proper environment to survive or even grow during or after the salting process, as some authors have already reported [52,53]. However, according to the growth limits of *S. aureus* stated by Valero et al. [54], the conditions of ripening and the characteristics of Pecorino Romano cheese: ripening temperature (10–12 °C), constant low pH (~5.3), and decreasing of a_w from 0.97 to 0.89 in the first 90 days of ripening, were enough severe to prevent proliferation and also to reduce the survival of *S. aureus* as shown by the results (<1 $\log_{10}$ CFU g^{-1}). These findings were in agreement with those observed by Pexara et al. [45] in Feta cheese with a salt content close to that of Pecorino Romano cheese.

Unlike the other bacteria tested in this study, the pathogenicity of *S. aureus* is due to the possible production of different enterotoxins in contaminated food. A level above 10^5 CFU mL^{-1} or g^{-1} of *S. aureus* can produce an infective enterotoxins concentration of about 1 μg [55]. Therefore, Regulation (EC) No 2073/2005 sets the limit in cheeses for coagulase-positive staphylococci (including *S. aureus*) at 10^5 CFU g^{-1} [51]. Above this limit, the presence of staphylococcal enterotoxins is expected, and there is a need to perform assays for enterotoxins detection, which must be absent in 25 g. However, staphylococcal enterotoxins production seems to be strongly limited in cheeses by the antagonistic effect of lactic acid bacteria. Microbial competition together with unfavorable environmental conditions, such as low pH, limits the growth of *S. aureus* and strongly downregulate virulence genes expression [52,53,56]. In particular, a rapid acidification in the first 6 h of cheese-making, as reported in our study (Figure 2), seems to be critical for *S. aureus* growth and enterotoxins production [46]. Furthermore, in our work, we simulated a worst-case contamination scenarios by employing unrealistically high starting *S. aureus* levels and despite this, the cheese-making conditions prevented the proliferation of *S. aureus*. Future studies will be carried out to examine the presence of staphylococcal enterotoxins in Pecorino Romano cheese produced from milk contaminated with *S. aureus*.

3.3. Pathogenic Bacteria Counts in TM Cheese-Making Trials

In Table 3 are reported the pathogens counts in milk (before and after thermization). Thermization is a sub-pasteurization heat treatment usually performed at 57–68 °C for 10–20 s to reduce the number of spoilage bacteria, with minimum collateral heat damage to milk components and thus provide a suitable environment for the growth of the added starter cultures [57]. Unlike the pasteurization, the thermization process is not able to ensure the complete inactivation of pathogenic microorganisms but, especially if they are present in high initial levels, it can only reduce their number, as confirmed by the pathogens counts in thermized milk shown in Table 3. In our study, we used a batch thermization at 65 °C, without resting at the set temperature. All the pathogens showed a reduction of around 3 $\log_{10}$ after thermization, starting from a bacterial count of about 6 $\log_{10}$ CFU mL^{-1} (Table 3). These rates of thermal inactivation are close to those reported by other authors, although a comparison with the time-temperature conditions reported in our work is not easy [58,59]. These studies highlight the heterogeneity in thermal resistance not only in different microbial groups but also between different strains of the same species. In particular, different strains of *E. coli* and *S. aureus* showed varying degrees of heat resistance, while *L. monocytogenes* and *Salmonella* spp. strains displayed a more uniform thermal susceptibility [59]. Hence, it is important to underline that a multi-strain inoculum should be used to represent the behavior of a given pathogen, as done in our work. The data reported in Table 3 show that counts levels in thermized milk did not differ significantly among the four pathogenic microorganisms, which indicate similar thermal tolerance under the conditions applied in the present study. However, we cannot exclude that some of the strains used in our study were more susceptible toward heat than others.

After 1 day from the production, *L. monocytogenes*, *E. coli* O157:H7, and *Salmonella* spp. dropped below the detection limit and were no longer detectable up to 150 days, even after an enrichment procedure, as shown in Table 4. *S. aureus* counts were also below the detection limit for the direct plating enumeration method ($<1 \log_{10}$ CFU g^{-1}) from 1 day up to the end of the ripening period. As already reported for RM trials, it was not possible to assess exactly the amount of the pathogens decrease related to the bacteria loss through the whey separation compared to the pathogens counts reduction due to the several environmental hurdles during the first stage of cheese-making. The latter may have a greater impact on TM trials, especially as a consequence of milk thermization, a second less severe thermal stress (curd cooking), and the faster and more intense acidification. In fact, after molding, the curd had a lower average pH value in TM trials (6.24) compared to RM trials (6.41). This difference was kept during the drainage process (Figure 2). Besides, as reported in Table 4, the microbial counts in RM and TM cheeses were significantly affected by milk thermization treatment ($p < 0.001$), as well as ripening time ($p < 0.001$). Furthermore, the interaction between these factors was also significant for pathogens counts ($p < 0.001$).

The introduction of the thermization in Pecorino Romano cheese-making is a further obstacle to the possible growth and survival of pathogens. In addition to other hurdles (curd cooking and acidification), it makes the cheese microbiologically safe before the salting process. In the case of RM cheeses, it is instead necessary to wait until the end of the salting process (90 days) to consider safety of the product. In any case, this happens 60 days before the mandatory minimum period for marketing PDO Pecorino Romano cheese (150 days).

Hence, our study shows that the production process of Pecorino Romano cheese both from raw and thermized milk is enough restrictive for the survival of the pathogens we have examined, except for *S. aureus* which may be still present in such low levels that it does not represent a hazard to the consumers. Currently, thermization is practiced by all producers of PDO Pecorino Romano cheese since this treatment has the advantage in standardizing the microbiological properties of the cheese milk by reducing the total microbial count and containing the undesirable bacteria, to obtain a constant quality of the product. This is a key element for a product aimed at an international market, such as Pecorino Romano cheese.

4. Conclusions

Cheeses made from unpasteurized milk are known to pose a health risk to the consumers. According to PDO specifications, Pecorino Romano cheese can be produced from raw or thermized whole sheep milk. In this study, Pecorino Romano cheese was produced from unpasteurized (raw and thermized) milk inoculated with *L. monocytogenes*, *Salmonella* spp., *E. coli* O157:H7, or *S. aureus*, in high initial loads (10^6 CFU mL^{-1}). The results demonstrated that the cheese-making process may ensure the microbiological safety of this cheese after the minimum ripening period (150 days) established by the PDO specifications. In particular, the obtained cheese was free from microbiological hazard after 90 days and 1 day, when manufactured from raw and thermized milk, respectively. These results could encourage cheese manufacturers to diversify Pecorino Romano cheese by using raw milk, which is currently almost not used, for its production, to provide the market with a product having a more intense flavor than that made from thermized milk. This is the first preliminary work to evaluate the survival of selected pathogenic bacteria during Pecorino Romano cheese ripening. Further studies are necessary to investigate the behavior of the pathogens in the early stages of manufacturing in order to understand more in depth on how their survival is hampered by different factors. The eventual production of staphylococcal enterotoxins will also be tested in subsequent investigations.

Author Contributions: Conceptualization, A.P. and A.F.; methodology, A.P., M.A. and A.F.; validation, G.L., R.M., M.P., M.A., A.F., and A.P.; formal analysis, G.L., R.M., and M.P.; investigation, G.L., R.M., and M.P.; resources, A.P., M.A., and A.F.; writing—original draft preparation, G.L., R.M., M.A., and M.P.; writing—review and editing, A.P. and A.F.; supervision A.P. and A.F. All authors have read and agreed to the published version of the manuscript.

Funding: This research received no external funding.

Acknowledgments: The technical staff of both the Technology and Chemistry sectors of Agris Sardegna is gratefully acknowledged.

Conflicts of Interest: The authors declare no conflict of interest.

References

1. Colonna, A.; Durham, C.; Meunier-Goddik, L. Factors affecting consumers' preferences for and purchasing decisions regarding pasteurized and raw milk specialty cheeses. *J. Dairy Sci.* **2011**, *94*, 5217–5226. [CrossRef] [PubMed]

2. Montel, M.; Buchin, S.; Mallet, A.; Delbes-Paus, C.; Vuitton, D.A.; Desmasures, N.; Berthier, F. Traditional cheeses: Rich and diverse microbiota with associated benefits. *Int. J. Food Microbiol.* **2014**, *177*, 136–154. [CrossRef] [PubMed]

3. Yoon, Y.; Lee, S.; Choi, K.-H. Microbial benefits and risks of raw milk cheese. *Food Control* **2016**, *63*, 201–215. [CrossRef]

4. Kousta, M.; Mataragas, M.; Skandamis, P.; Drosinos, E.H. Prevalence and sources of cheese contamination with pathogens at farm and processing levels. *Food Control* **2010**, *21*, 805–815. [CrossRef]

5. Verraes, C.; Vlaemynck, G.; Van Weyenberg, S.; De Zutter, L.; Daube, G.; Sindic, M.; Uyttendaelec, M.; Herman, L. A review of the microbiological hazards of dairy products made from raw milk. *Int. Dairy J.* **2015**, *50*, 32–44. [CrossRef]

6. Langer, A.J.; Ayers, T.; Grass, J.; Lynch, M.; Angulo, F.J.; Mahon, B.E. Nonpasteurized Dairy Products, Disease Outbreaks, and State Laws—United States, 1993–2006. *Emerg. Infect. Dis.* **2012**, *18*, 385–391. [CrossRef]

7. Costard, S.; Espejo, L.; Groenendaal, H.; Zagmutt, F.J. Outbreak-related disease burden associated with consumption of unpasteurized cow's milk and cheese, United States, 2009–2014. *Emerg. Infect. Dis.* **2017**, *23*, 957–964. [CrossRef]

8. Rocourt, J. Foodborne diseases: Foodborne diseases and vulnerable groups. In *Encyclopedia of Food Safety*, 1st ed.; Motarjemi, Y., Ed.; Academic Press: Waltham, MA, USA, 2014; Volume 1, pp. 323–331. ISBN 978-0-12-378613-5.

9. Ioanna, F.; Quaglia, N.C.; Storelli, M.M.; Castiglia, D.; Goffredo, E.; Storelli, A.; De Rosa, M.; Normanno, G.; Jambrenghi, A.C.; Dambrosio, A. Survival of *Escherichia coli* O157:H7 during the manufacture and ripening of *Cacioricotta* goat cheese. *Food Microbiol.* **2018**, *70*, 200–205. [CrossRef]

10. Thakur, M.; Asrani, R.K.; Patial, V. Listeria monocytogenes: A Food-Borne Pathogen. In *Foodborne Diseases*, 1st ed.; Holban, A.M., Grumezescu, A.M., Eds.; Academic Press: London, UK, 2018; Volume 15, pp. 157–192. ISBN 978-0-12-811444-5.

11. Arqués, J.L.; Rodríguez, E.; Langa, S.; Landete, J.M.; Medina, M. Antimicrobial Activity of Lactic Acid Bacteria in Dairy Products and Gut: Effect on Pathogens. *BioMed Res. Int.* **2015**, *2015*, 584183. [CrossRef]

12. Melero, B.; Stessl, B.; Manso, B.; Wagner, M.; Esteban-Carbonero, Ó.J.; Hernández, M.; Rovira, J.; Rodriguez-Lázaro, D. *Listeria monocytogenes* colonization in a newly established dairy processing facility. *Int. J. Food Microbiol.* **2019**, *289*, 64–71. [CrossRef]

13. Castro, R.D.; Pedroso, S.H.S.P.; Sandes, S.H.C.; Silva, G.O.; Luiz, K.C.M.; Dias, R.S.; Filho, R.A.T.; Figueiredo, H.C.P.; Santos, S.G.; Nunes, A.C.; et al. Virulence factors and antimicrobial resistance of *Staphylococcus aureus* isolated from the production process of Minas artisanal cheese from the region of Campo das Vertentes, Brazil. *J. Dairy Sci.* **2020**, *103*, 2098–2110. [CrossRef] [PubMed]

14. Lahou, E.; Uyttendaele, M. Growth potential of *Listeria monocytogenes* in soft, semi-soft and semi-hard artisanal cheeses after post-processing contamination in deli retail establishments. *Food Control* **2017**, *76*, 13–23. [CrossRef]

15. Castañeda-Ruelas, G.M.; Soto-Beltrán, M.; Chaidez, C. Detecting Sources of *Staphylococcus aureus* in One Small-Scale Cheese Plant in Northwestern Mexico. *J. Food Saf.* **2016**, *37*, e12290. [CrossRef]

16. Normanno, G.; Spinelli, E.; Caruso, M.; Fraccalvieri, R.; Capozzi, L.; Barlaam, A.; Parisi, A. Occurrence and characteristics of methicillin-resistant *Staphylococcus aureus* (MRSA) in buffalo bulk tank milk and the farm workers in Italy. *Food Microbiol.* **2020**, *91*, 103509.

17. D'Amico, D.J.; Donelly, C.W. Growth and Survival of Microbial Pathogens in Cheese. In *Cheese*, 4th ed.; McSweeney, P.L.H., Fox, P.F., Cotter, P.D., Everett, D.W., Eds.; Academic Press: San Diego, CA, USA, 2017; Volume 1, pp. 573–594. ISBN 978-0-12-417012-4.

18. Alvarez-Ordóñez, A.; Broussolle, V.; Colin, P.; Nguyen-The, C.; Prieto, M. The adaptive response of bacterial foodborne pathogens in the environment, host and food: Implications for food safety. *Int. J. Food Microbiol.* **2015**, *213*, 99–109. [CrossRef]

19. Melo, J.; Andrew, P.W.; Faleiro, M.L. *Listeria monocytogenes* in cheese and the dairy environment remains a food safety challenge: The role of stress responses. *Food Res. Int.* **2015**, *67*, 75–90. [CrossRef]

20. Ye, B.; He, S.; Zhou, X.; Cui, Y.; Zhou, M.; Shi, X. Response to Acid Adaptation in *Salmonella enterica* Serovar Enteritidis. *J. Food Sci.* **2019**, *84*, 599–605. [CrossRef]

21. Bellio, A.; Bianchi, D.M.; Vitale, N.; Vernetti, L.; Gallina, S.; Decastelli, L. Behavior of *Escherichia coli* O157:H7 during the manufacture and ripening of Fontina Protected Designation of Origin cheese. *J. Dairy Sci.* **2018**, *101*, 4962–4970. [CrossRef]

22. Dalzini, E.; Cosciani-Cunico, E.; Monastero, P.; Bernini, V.; Neviani, E.; Bellio, A.; Decastelli, L.; Losio, M.N.; Daminelli, P.; Varisco, G. *Listeria monocytogenes* in Gorgonzola cheese: Study of the behaviour throughout the process and growth prediction during shelf life. *Int. J. Food Microbiol.* **2017**, *262*, 71–79. [CrossRef]

23. Peng, S.; Hoffmann, W.; Bockelmann, W.; Hummerjohann, J.; Stephan, R.; Hammer, P. Fate of Shiga toxin-producing and generic *Escherichia coli* during production and ripening of semihard raw milk cheese. *J. Dairy Sci.* **2013**, *96*, 815–823. [CrossRef]

24. D'Amico, D.J.; Druart, M.J.; Donnelly, C.W. Behavior of *Escherichia coli* O157:H7 during the manufacture and aging of Gouda and stirred-curd cheddar cheeses manufactured from raw milk. *J. Food Prot.* **2010**, *73*, 2217–2224. [CrossRef] [PubMed]

25. D'Amico, D.J.; Druart, M.J.; Donnelly, C.W. Comparing the Behavior of Multidrug-Resistant and Pansusceptible Salmonella during the Production and Aging of a Gouda Cheese Manufactured from Raw Milk. *J. Food Prot.* **2014**, *77*, 903–913. [CrossRef] [PubMed]

26. Chatelard-Chauvin, C.; Pelissier, F.; Hulin, S.; Montel, M.C. 2015. Behaviour of *Listeria monocytogenes* in raw milk Cantal type cheeses during cheese making, ripening and storage in different packaging conditions. *Food Control* **2015**, *54*, 53–65. [CrossRef]

27. Consorzio per la Tutela del Formaggio Pecorino Romano. Specification and Regulations. Available online: https://www.pecorinoromano.com/en/pecorino-romano/specification-and-regulations (accessed on 27 May 2020).

28. Addis, M.; Fiori, M.; Riu, G.; Pes, M.; Salvatore, E.; Pirisi, A. Physico-chemical and nutritional characteristics of PDO Pecorino Romano cheese: Seasonal variation. *Small Rumin. Res.* **2015**, *126*, 73–79. [CrossRef]

29. CLAL. Italy: Export Pecorino e Fiore Sardo. 2020. Available online: https://www.clal.it/?section=impexpistat&cod=04069063&mov=E (accessed on 16 June 2020).

30. Meunier-Goddik, L.; Waite-Cusic, J. Consumers Acceptance of Raw Milk and its Products. In *Raw Milk*, 1st ed.; Nero, L.A., De Carvalho, A.F., Eds.; Academic Press: San Diego, CA, USA, 2019; pp. 311–350. ISBN 978-0-12-810530-6.

31. Brooks, J.C.; Martinez, B.; Stratton, J.; Bianchini, A.; Krokstrom, R.; Hutkins, R. Survey of raw milk cheeses for microbiological quality and prevalence of foodborne pathogens. *Food Microbiol.* **2012**, *31*, 154–158. [CrossRef]

32. *ISO 5534:2004. Cheese and Processed Cheese—Determination of the Total Solids content*; International Organization for Standardization: Geneva, Switzerland, 2004.

33. *ISO 5943:2006. Cheese and Processed Cheese products—Determination of Chloride Content—Potentiometric Titration Method*; International Organization for Standardization: Geneva, Switzerland, 2006.

34. *ISO 11290-1:2017. Microbiology of the Food Chain—Horizontal Method for the Detection and Enumeration of Listeria Monocytogenes and of Listeria spp.—Part 1: Detection Method*; International Organization for Standardization: Geneva, Switzerland, 2017.

35. *ISO 11290-2:2017. Microbiology of the Food Chain—Horizontal Method for the Detection and Enumeration of Listeria Monocytogenes and of Listeria spp.—Part 2: Enumeration Method*; International Organization for Standardization: Geneva, Switzerland, 2017.

36. *ISO 6888-2:2004. Microbiology of Food and Animal Feeding Stuffs—Horizontal Method for the Enumeration of Coagulase-Positive Staphylococci (Staphylococcus aureus and Other Species)—Technique Using Rabbit Plasma Fibrinogen Agar Medium*; International Organization for Standardization: Geneva, Switzerland, 2004.

37. *ISO 16654:2001/A1:2017. Microbiology of Food and Animal Feeding Stuffs—Horizontal Method for the Detection of Escherichia Coli O157—Amendment 1: Annex B: Result of Interlaboratory Studies*; International Organization for Standardization: Geneva, Switzerland, 2017.

38. *ISO 6579-1:2017. Microbiology of the Food Chain—Horizontal Method for the Detection, Enumeration and Serotyping of Salmonella—Part 1: Detection of Salmonella Spp.*; International Organization for Standardization: Geneva, Switzerland, 2017.

39. Xiong, L.; Li, C.; Boeren, S.; Vervoort, J.; Hettinga, K. Effect of heat treatment on bacteriostatic activity and protein profile of bovine whey proteins. *Food Res. Int.* **2020**, *127*, 108688. [CrossRef]

40. McSweeney, P.L.H.; Fox, P.F.; Ciocia, F. Metabolism of Residual Lactose and of Lactate and Citrate. In *Cheese: Chemistry, Physics and Microbiology*, 4th ed.; Mc Sweeney, P., Fox, P., Cotter, P., Everett, D., Eds.; Elsevier: Amsterdam, The Netherlands, 2017; pp. 411–421. ISBN 978-0-12-417012-4.

41. McSweeney, P.L.H. Biochemistry of cheese ripening: Introduction and overview. In *Cheese: Chemistry, Physics and Microbiology*, 4th ed.; McSweeney, P., Fox, P., Cotter, P., Everett, D., Eds.; Elsevier: Amsterdam, The Netherlands, 2017; pp. 379–387. ISBN 978-0-12-417012-4.

42. Addis, M.; Piredda, G.; Pes, M.; Di Salvo, R.; Scintu, M.F.; Pirisi, A. Effect of the use of three different lamb paste rennets on lipolysis of the PDO Pecorino Romano Cheese. *Int. Dairy J.* **2005**, *15*, 563–569. [CrossRef]

43. Peng, S.; Tasara, T.; Hummerjohann, J.; Stephan, R. An overview of molecular stress response mechanisms in *Escherichia coli* contributing to survival of Shiga toxin-producing *Escherichia coli* during raw milk cheese production. *J. Food Prot.* **2011**, *74*, 849–864. [CrossRef]

44. Modi, R.; Hirvi, Y.; Hill, A.; Griffiths, M.W. Effect of phage on survival of *Salmonella* Enteritidis during manufacture and storage of cheddar cheese made from raw and pasteurized milk. *J. Food Protect.* **2001**, *64*, 927–933. [CrossRef]

45. Pexara, A.; Solomakos, N.; Sergelidis, D.; Govaris, A. Fate of enterotoxigenic *Staphylococcus aureus* and staphylococcal enterotoxins in Feta and Galotyri cheeses. *J. Dairy Res.* **2012**, *79*, 405–413. [CrossRef]

46. Delbes, C.; Alomar, J.; Chougui, N.; Martin, J.F.; Montel, M.C. *Staphylococcus aureus* growth and enterotoxin production during the manufacture of uncooked, semihard cheese from cows' raw milk. *J. Food Prot.* **2006**, *69*, 2161–2167. [CrossRef] [PubMed]

47. Farrokh, C.; Jordan, K.; Auvray, F.; Glass, K.; Oppegaard, H.; Raynaud, S.; Thevenot, D.; Condron, R.; De Reu, K.; Govaris, A.; et al. Review of Shiga-toxin-producing *Escherichia coli* (STEC) and their significance in dairy production. *Int. J. Food Microbiol.* **2013**, *162*, 190–212. [CrossRef]

48. Calicioglu, M.; Dikici, A. Survival and Acid Adaptation Ability of *Salmonella* during Processing and Ripening of *Savak Tulumi* Cheese. *J. Anim. Vet. Adv.* **2009**, *8*, 1124–1130.

49. Papadopoulou, C.; Maipa, V.; Dimitriou, D.; Pappas, C.; Voutsinas, L.; Malatou, H. Behavior of *Salmonella enteritidis* During the Manufacture, Ripening, and Storage of Feta Cheese Made from Unpasteurized Ewe's Milk. *J. Food Prot.* **1993**, *56*, 24–28. [CrossRef] [PubMed]

50. Wusimanjiang, P.; Ozturk, M.; Ayhan, Z.; Mehmetoglu, A.Ç. Effect of salt concentration on acid- and salt-adapted *Escherichia coli* O157:H7 and *Listeria monocytogenes* in recombined nonfat cast cheese. *J. Food Process Pres.* **2019**, *43*, e14208. [CrossRef]

51. Consolidated Text: Commission Regulation (EC) No 2073/2005 on Microbiological Criteria for Food Stuffs. Available online: https://eur-lex.europa.eu/legal-content/EN/TXT/?uri=CELEX%3A02005R2073-20200308 (accessed on 30 October 2020).

52. Al-Nabulsi, A.A.; Osaili, T.M.; AbuNaser, R.A.; Olaimat, A.N.; Ayyash, M.; Al-Holy, M.A.; Kadora, K.M.; Holley, R.A. Factors affecting the viability of *Staphylococcus aureus* and production of enterotoxin during processing and storage of white-brined cheese. *J. Dairy Sci.* **2020**, *103*, 6869–6881. [CrossRef]

53. Mohammadi, K.; Hanifian, S. Growth and enterotoxin production of *Staphylococcus aureus* in Iranian ultra-filtered white cheese. *Int. J. Dairy Technol.* **2015**, *68*, 111–117. [CrossRef]

54. Valero, A.; Pérez-Rodríguez, F.; Carrasco, E.; Fuentes-Alventosa, J.M.; García-Gimeno, R.; Zurera, G. Modelling the growth boundaries of *Staphylococcus aureus*: Effect of temperature, pH and water activity. *Int. J. Food Microbiol.* **2009**, *133*, 186–194. [CrossRef]

55. Ahmed, A.A.-H.; Maharik, N.M.S.; Valero, A.; Kamal, S.M. Incidence of enterotoxigenic *Staphylococcus aureus* in milk and Egyptian artisanal dairy products. *Food Control* **2019**, *104*, 20–27. [CrossRef]

56. Viçosa, G.N.; Botelho, C.V.; Botta, C.; Bertolino, M.; de Carvalho, A.F.; Nero, L.A.; Cocolin, L. Impact of co-cultivation with *Enterococcus faecalis* over growth, enterotoxin production and gene expression of *Staphylococcus aureus* in broth and fresh cheeses. *Int. J. Food Microbiol.* **2019**, *308*, 108291. [CrossRef]

57. Rukke, E.O.; Sørhaug, T.; Stepaniak, L. Heat Treatment of Milk|Thermization of Milk. In *Encyclopedia of Dairy Sciences*, 2nd ed.; Fuquay, J.W., Ed.; Academic Press: San Diego, CA, USA, 2011; pp. 693–698. ISBN 978-0-12-374407-4.

58. Gabriel, A.A.; Bayaga, C.L.T.; Magallanes, E.A.; Aba, R.P.M.; Tanguilig, K.M.N. Fates of pathogenic bacteria in time-temperature-abused and Holder-pasteurized human donor-, infant formula-, and full cream cow's milk. *Food Microbiol.* **2020**, *89*, 103450. [CrossRef] [PubMed]

59. Pearce, L.E.; Smythe, B.W.; Crawford, R.A.; Oakley, E.; Hathaway, S.C.; Shepherd, J.M. Pasteurization of milk: The heat inactivation kinetics of milk-borne dairy pathogens under commercial-type conditions of turbulent flow. *J. Dairy Sci.* **2012**, *95*, 20–35. [CrossRef] [PubMed]

Publisher's Note: MDPI stays neutral with regard to jurisdictional claims in published maps and institutional affiliations.

Article

Functional Odd- and Branched-Chain Fatty Acid in Sheep and Goat Milk and Cheeses

Anna Nudda, Fabio Correddu *, Alberto Cesarani, Giuseppe Pulina and Gianni Battacone

Dipartimento di Agraria, Sezione di Scienze Zootecniche, University of Sassari, viale Italia, 39, 07100 Sassari, Italy; anudda@uniss.it (A.N.); acesarani@uniss.it (A.C.); gpulina@uniss.it (G.P.); battacon@uniss.it (G.B.)
* Correspondence: fcorreddu@uniss.it; Tel.: +39-079-229-308

Citation: Nudda, A.; Correddu, F.; Cesarani, A.; Pulina, G.; Battacone, G. Functional Odd- and Branched-Chain Fatty Acid in Sheep and Goat Milk and Cheeses. *Dairy* **2021**, *2*, 79–89. https://doi.org/10.3390/dairy2010008

Received: 9 December 2020
Accepted: 27 January 2021
Published: 5 February 2021

Publisher's Note: MDPI stays neutral with regard to jurisdictional claims in published maps and institutional affiliations.

Abstract: The inverse association between the groups of odd-chain (OCFA) and branched-chain (BCFA) and the development of diseases in humans have generated interest in the scientific community. In experiment 1, the extent of the passage of odd- and branched-chain fatty acids (OBCFA) from milk fat to fresh cheese fat was studied in sheep and goats. Milk collected in two milk processing plants in west Sardinia (Italy) was sampled every 2 weeks during spring (March, April and May). In addition, a survey was carried out to evaluate the seasonal variation of the OBCFA concentrations in sheep and goats' cheeses during all lactation period from January to June. Furthermore, to assess the main differences among the sheep and goat cheese, principal component analysis (PCA) was applied to cheese fatty acids (FA) profile. Concentrations of OBCFA in fresh cheese fat of both species were strongly related to the FA content in the unprocessed raw milk. The average contents of OBCFA were 4.12 and 4.13 mg/100 mg of FA in sheep milk and cheese, respectively, and 3.12 and 3.17 mg/100 mg of FA in goat milk and cheese, respectively. The OBCFA concentration did no differed between milk and cheese in any species. The content of OBCFA was significantly higher in sheep than goats' dairy products. The OBCFA composition of the cheese was markedly affected by the period of sampling in both species: odd and branched FA concentrations increased from March to June. The seasonal changes of OBCFA in dairy products were likely connected to variations in the quality of the diet. The PCA confirmed the higher nutritional quality of sheep cheese for beneficial FA, including OBCFA compared to the goat one, and the importance of the period of sampling in the definition of the fatty acids profile.

Keywords: sheep and goat milk; cheese; odd and branched chain fatty acids

1. Introduction

The fatty acid (FA) composition of dairy products has assumed considerable interest in consumers from a nutritional and healthy point of view. Indeed, specific FA of dairy products can affect human health and can have an important role in the prevention of metabolic diseases. This increasing attention is also demonstrated by recent studies that investigated the feasibility of improving milk FA profile in sheep and goats though breeding schemes [1–3]. A significant amount of attention has been focused in the last decade on polyunsaturated fatty acid of the omega3 family (PUFA n-3) and conjugated linoleic acid (CLA) concentrations of dairy products. Moreover, the groups of odd-chain fatty acids (OCFA) and branched-chain fatty acids (BCFA), long neglected due to their low incidence on the total amount of FA, have sparked interest in the scientific community due to an inverse relationship with the development of human diseases.

In particular, the group of BCFA comprises mainly saturated fatty acids characterized by the presence of one or more methyl groups in the iso or anteiso position. Such molecules represent the main FAs in some microorganisms (e.g., Bacilli and Lactobacilli), and they can be also observed in mammal tissues. In laboratory animals these FA evidenced anti-inflammatory properties [4], reduced the incidence of necrotizing enterocolitis and altered the ecology of the gastrointestinal microorganisms in a neonatal rat model [5].

The OCFA, which include pentadecanoic (C15:0) and heptadecanoic (C17:0) acids and their isomeric forms, are mostly derived from ruminal bacteria cell wall and then partitioned firstly to organs and tissues and, therefore, detectable in ruminant-derived foods. Thus, dairy foods or ruminant fats represent the major dietary source of OCFA for humans.

Moreover, the milk concentrations of C15:0 and C17:0 are considered biomarkers of rumen microbial fermentation and microbial de novo lipogenesis [6]. In addition, the mammary gland plays a role in their synthesis by using propionate [7], and the subcutaneous adipose tissue and, therefore, after their mobilization, can be incorporated into milk fat [8].

Humans are not able to synthetize C15:0 and C17:0 so they could be considered valuable markers of dairy fat intake [9,10]. The role of these FA in human health has recently been reinforced in view of new biological and nutritional observations. The C15:0 and C17:0 have been inversely associated with cardiovascular disease [11–14] and incidence of type 2 diabetes [15–18]. The direct role of C15:0 in attenuating pro-inflammatory state, cytotoxicity in human cell lines, anemia, and dyslipidemia lowering glucose and cholesterol in vivo has been recently evidenced [19].

Among the different factors influencing the OBCFA concentration in milk and dairy products, animal diet is the main one. Indeed, the diet of animals can modulate microbial growth in the rumen, de novo microbial FA synthesis and the uptake of blood circulating FA by the mammary gland. Variations in milk OBCFA have been observed in response to lipid supplementation [20–22], change in forage to concentrate ratio [23] and use of by-products rich in bioactive compounds [24]. Changes in the OBCFA concentration in milk has been also observed with lactation stage [25] and with energy balance of animals [26], probably as a consequence of the fat mobilization from adipose tissue and the extent of mammary uptake for milk fat synthesis [27].

In Mediterranean areas, almost all sheep milk is processed into cheese, and milk fatty acid profile has important effects on cheese fat quality under nutritional point of view. The total transfer from milk into cheese of fatty acids of nutritional interest, as PUFA n-3 and CLA, has been reported in sheep [28], but effects of cheese-making technology has been also evidenced [29]. However, the mechanisms influencing the relationship between OBCFA concentrations in unprocessed milk and the derived dairy products are still unclear. The presence of FA in the FA composition of different microorganism can be of significant importance for the dairy industry where different microbial cultures, as starters, are widely used for making different products. The consequence could be the change of the native OCFA and BCFA milk concentrations after processing in cheese or other dairy products.

The aims of this study were (a) to evaluate the extent of OBCFA transfer from ovine and caprine milk to cheese (Experiment 1); (b) to describe the seasonal variation of OBCFA in sheep and goat cheeses (Experiment 2). Both were sequentially investigated with two independent surveys.

2. Materials and Methods

In Experiment 1, bulk tank milk of Sarda breed sheep and Sardinian goats and cheeses were collected from processing plants in North Sardinia for sheep (n. 2) and in west Sardinia for goats (n. 2). Samples were taken every two weeks during spring (March, April and May) for a total of 12 milk samples per species. The transfer of OBCFA from milk to fresh cheeses after 24 h from processing was evaluated.

In Experiment 2, mild sheep and goat cheeses, with a ripening time of about 20–30 days, were sampled monthly from January to July from nine sheep cheese-making plants located in different areas of Sardinia (Guspini, Nurri, Marrubiu, Macomer, Dorgali, Oliena, Onifai, Thiesi, Berchidda), five plants of which produced also goat cheese. The samples collection started in January, which corresponds to the begin of lactation both in sheep and goats reared in Sardinia's breeding system. The goats cheese samples were from four lots/month from January to April and three different lots/month from May to July for a total of 25 goat cheese samples. The sheep cheese samples were Pecorino Sardo PDO type from

8 lots/month from January to May, and from four different lots/month in June and July for a total of 48 sheep cheese samples.

The determination of FA concentrations in milk and cheese samples was carried out by a gas chromatographic method, as previously described [20]. Briefly, after the lipid extraction, a base-catalyzed trans-esterification FIL-IDF standard procedure [30] was used to prepare FA methyl esters (FAME). Identification of individual FAME was allowed by comparing their retention time with that of a series of analytical standards. Those included the Supelco 37 component FAME MIX (Supelco, Bellefonte, PA, USA), the GLC-110 MIX (Matreya Inc. Pleasant Gap, PA, USA) and some individual BCFA (Matreya Inc. Pleasant Gap, PA, USA). Identification of OBCFA was also supported by the consultation of previous studies [31,32]. The FA concentration was reported as mg/100 mg of total FAME.

Using R software (R Core Team, 2020), general linear model (GLM) was applied to data on OBCFA and others main nutritional FA (R Core Team, 2020). The considered fixed effects were species (sheep and goat) and product type (milk and cheese) for experiment 1, whereas species and month for Experiment 2. Significative differences were declared at $p < 0.05$.

Principal component analysis (PCA) was applied on sixty-three FA profile of sheep and goats' cheeses using *prcomp* R function. A total of 42 FA was considered and, therefore, 42 principal components were extracted. Before PCA, all considered FA were scaled to ensure unit variance.

3. Results and Discussion

Means ($\pm$SE) of fat concentration in milk and cheese of sheep and goat, respectively were 6.16% ($\pm$0.26) and 5.42% ($\pm$0.03), 28.81% ($\pm$0.65) and 21.17% ($\pm$0.54).

3.1. Experiment 1: Fatty Acid Transfer from Milk to Cheese

FA composition in milk and cheese for sheep and goats is reported in Table 1. In total, 10 OBCFA were identified in both types of milk (Table 1), including 3 OCFA (C13:0, C15:0 and C17:0), 4 isoBCFA (isoC14:0, isoC15:0, isoC16:0, and isoC17:0) and 3 anteisoBCFA (anteisoC13:0, anteisoC15:0 and anteisoC17:0).

Among them, C15:0 and C17:0 were the most abundant fatty acids, in agreement with other research on sheep [33], goats and cows [34,35]; these OCFA accounted for 29% and 16% of the total concentration of OBCFA, respectively. In our study, C13:0 represents only 2.3%, value that agrees with the observations in dairy cows in which it represents 2% of OCFA [34] or it has not been found [36], but lower than values previously reported in sheep's milk, where accounted for 6.3% of total OCFA [33].

Concentrations of individual OBCFA in the fat of fresh cheeses did not differ from that of raw unprocessed milk ($p > 0.05$) in both species: it means that overall, OBCFA content of milk is not significantly modified by cheese making processes, both in sheep and goats.

In sheep milk, the content of total OBCFA is higher than that observed in goat milk, due to a higher content of isoBCFA form of C15:0, C16:0 and C17:0, and anteiso C15:0 and C17:0 ($p < 0.05$). No differences among species have been observed only for isoC13:0 and isoC14:0. The differences between species could be related to their different feeding regimen, as OBCFA in milk may change according to the bacterial population present in the rumen and, particularly, because of the differences in their relative abundance. Differences in OBCFA concentrations between these the species can occur also because of different extents of de novo synthesis of these FA in the mammary gland, differential efficiency of intestinal absorption, and differential storage of some OBCFA in adipose tissue, followed by their release during periods of fat mobilization.

The C17:1c9 was not detected in sheep and it was found in negligible amount in goat milk and fresh cheese. In the mammary gland, the delta-9 desaturase activity can promote the metabolization of some odd-chain fatty acids; only the conversion of C17:0 to C17:1cis9 has been found of quantitative importance in cows [37]. It has also been reported with a mean concentration of 0.2% in milk of Assaf ewes raised in controlled and intensive farming system [33]. Therefore, the lack of C17:1c9 in Sarda dairy ewes might be related to specific environmental condition related to the typical extensive breeding system of Sarda dairy ewes and Sardinian goats.

The total content of OBCFA in sheep milk is higher in this experiment than that reported for cows [36] and sheep (3.19 g/100 g of total FA) [33]. Differences could be related to different feeding strategies at farm level, as grazing pasture, or vegetable lipid supplementation, forage to concentrate ratio or other dietary factors, that deserve to be further investigated.

Table 1. Odd and branched chain fatty acid in milk and cheese of sheep and goat species.

	Sheep			Goat			*p*-Value	
	Milk	Cheese	SEM [1]	Milk	Cheese	SEM [1]	M*vs*C [2]	Species
Fat, %	6.15	28.81		5.42	21.17		*	*
Fatty acid [3]								
iso C13:0	0.015	0.022	0.002	0.018	0.017	0.004	ns	ns
iso C14:0	0.105	0.102	0.006	0.107	0.111	0.005	ns	ns
iso C15:0	0.288	0.278	0.016	0.233	0.236	0.023	ns	**
iso C16:0	0.288	0.290	0.006	0.239	0.253	0.018	ns	**
iso C17:0	0.420	0.406	0.009	0.159	0.170	0.015	ns	**
anteiso C13:0	0.050	0.052	0.010	0.066	0.073	0.014	ns	*
anteiso C15:0	0.560	0.546	0.010	0.374	0.392	0.043	ns	**
anteiso C17:0	0.495	0.502	0.008	0.381	0.399	0.031	ns	*
C13:0	0.095	0.080	0.011	0.070	0.072	0.009	ns	*
C15:0	1.178	1.214	0.052	0.880	0.893	0.073	ns	**
C17:0	0.668	0.656	0.029	0.787	0.749	0.031	ns	**
C17:1 cis9	0.000	0.000	0.000	0.007	0.002	0.007	ns	*
BCFA	2.22	2.20	0.037	1.45	1.51	0.128	ns	*
OCFA	2.00	2.00	0.075	1.65	1.67	0.083	ns	*
OBCFA	4.22	4.20	0.103	3.16	3.12	0.216	ns	*

[1] SEM = standard error of means; [2] M*vs*C = Milk vs. Cheese; [3] Fatty acids: expressed as mg/100 mg of FAME; ** $p \leq 0.01$, * $p < 0.05$, ns not significant $p \geq 0.05$.

3.2. Experiment 2: Seasonal Variation of Branched and Odd Chain FA

Seasonal variation of individual and group of branched and odd chain fatty acid has been investigated by collecting cheese samples throughout all lactation period of sheep and goats, that in Sardinia, as in all Mediterranean environmental conditions, occurs from January (start of lactation) to July (when animals are dry off).

The temporal evolution of these FA in cheese showed an increase of total OCFA and BCFA in both species ($p < 0.05$) across all the sampling period (Figure 1a,b); the highest contents of these FA were observed in the cheese produced with milk in advanced lactation, which corresponds also to the hot season. The increase of OBCFA across lactation has been previously reported in dairy cows [34].

Among individual OCFA, the content of odd C15:0 during lactation followed the same pattern of the total OCFA, being the main representative of this group of FA (Figure 2a), whereas C17:0 showed constant concentration during the early and mid-lactation, then increased at the end of lactation, in June and July for both species (Figure 2b). As far as the changes of individual BCFA during lactation are concerned, both anteisoC15:0 and anteisoC17:0 (Figure 2c,d) followed a similar pattern observed for total BCFA.

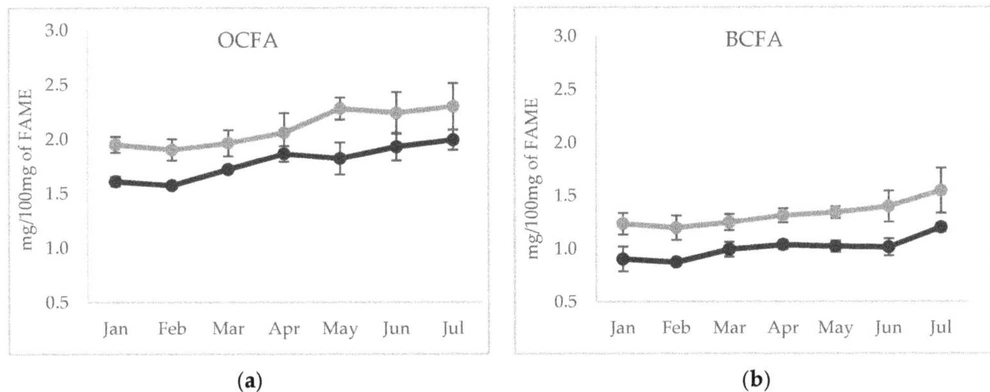

Figure 1. Temporal evolution of (**a**) odd chain fatty acids (FA) and (**b**) branched chain FA from January to July in cheeses from sheep (light grey line) and goats (dark grey line) milk.

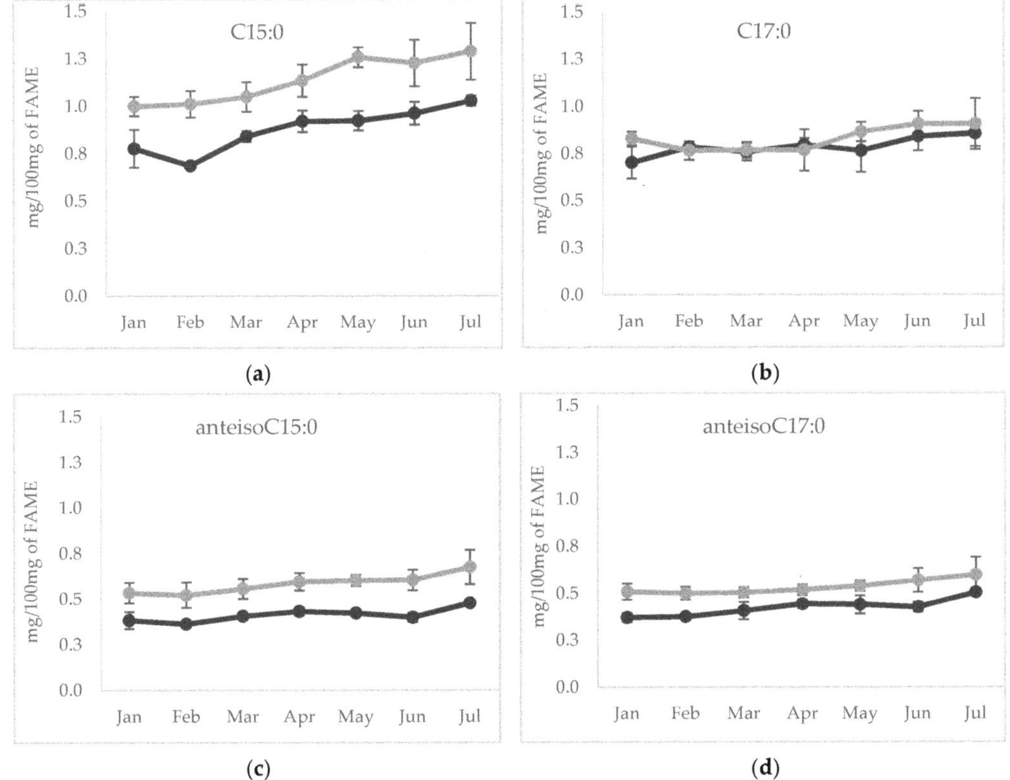

Figure 2. Temporal evolution of (**a**) C15:0, (**b**) C17:0, (**c**) anteisoC15:0) and (**d**) anteisoC17:0 from January to July in sheep (light grey line) and in goats (dark grey line) cheeses.

Changes in the animals' diet during the period under investigation, and in particular the progressive modification of pasture grazed by animals, can be the most probable explanation for the observed FA patterns. In fact, the typical lactation curve of the Sarda sheep breed follows the natural availability, both in quantitative and qualitative terms, of pastures [38]. As lactation progress, there is a worsening of nutritional quality of pasture characterized by increase of fiber and a reduction of protein and FA contents, especially alpha-linolenic acid (C18:3n3), which concentrations decrease in mature grass [39,40]. This is supported by evolution of C18:3in cheeses samples as lactation progress (data not reported), that confirmed the pattern of Sarda ewes previously observed [28].

Different mechanisms could be hypothesized to explain this pattern. One mechanism could be related to the decrease of polyunsaturated fatty acids (PUFA) in the pasture that could have reduced the toxic effect of unsaturated lipids on microbial growth [41,42], especially of cellulolytic bacteria [43]. This could increase the rumen microbial abundance, followed by higher rumen outflow of fatty acid of microbial origin. A second possible explanation could be the positive association between OBCFA and dietary fiber as with progress of lactation there is an increase of neutral detergent fiber (NDF), especially acid detergent lignin (ADL) in pasture. This is supported by previously observation in dairy goats where dietary NDF was found the most important factor of variation in lipid composition of bacteria; and in dairy cows where the proportion of odd- and branched-chain FA increased and those of even-chain saturated FA decreased with increasing forage [44].

The contents of almost all OBCFA in goat cheese differed from that of sheep, confirming the findings of the experiment 1. It is noteworthy that the goat's milk processed into cheese in this Mediterranean area comes from animals raised in extensive system, characterized by natural pastures rich also in shrubs and essences of the Mediterranean maquis, which contains tannins. These different pasture conditions among species could be a possible explanation of the different concentrations in OBCFA observed between sheep and goats [45].

In order to assess the overall differences among sheep and goat cheese in terms of FA composition, a multivariate statistical analysis was carried out on the detailed FA profile. In particular, PCA was used because it is a useful instrument to reduce the complexity of a multivariate space, such as that of FA of cheese, and has demonstrated to be able to separate samples (e.g., of different origin, dietary treatment, season of production, lactation stage), based exclusively on the FA composition [46–48]. An important application of PCA could be, as an example, the authentication of dairy products according to their lipid profile [49].

The first five principal components (PCs) explained the 80% of the total variance, with the two first PCs accounting for more than 50% (PC1 and PC2, 36% and 22%, respectively). The plot of the scores of the PC1 and PC2 (Figure 3a) allows the clustering of cheeses according to the season of milk production (PC1) or to their species of origin (PC2). Cheeses produced during winter–early spring season had positive scores for PC1, whereas those produced during late-spring–summer season had negative scores. Regarding the PC2, goat and sheep cheese had positive and negative scores, respectively. In the same plot, can be also observed that the Fa profile of sheep cheese was characterized by a larger variability compared to goat one, across the seasons of production.

The loadings of the first two PCs are plotted in Figure 3b. PC loadings can be used to describe and explain the main FA differences among the clusters of cheese observed in Figure 3a, i.e., the plot of the scores (cheese produced in different season or from milk of different species).

The original variables (i.e., single fatty acids) exhibiting the highest loading values for the 2 PCs (both negatives and positives) can be used to describe the main differences, in terms of FA, among samples belonging to the cheese of the two species and, also, produced in different seasons.

PC1 had high positive loadings for short chain FA (C4:0, C6:0, C8:0, C10:0), alpha linolenic acid, vaccenic and rumenic acid. High negative loadings for PC1 were observed for different long chain saturated FA (C20:0, C22:0, C24:0) and for the C18:1cis-9. This pattern was quite consistent with the milk (and cheese) FA variation during seasons, due to the worsening of pasture quality: in particular, the decrease of alpha-linolenic, vaccenic and rumenic acids in dairy products is correlated to the decrease in pasture of alpha-linolenic acid content from winter to summer. In addition, PC1 loadings can be also related to the lactation progress (high short chain FA in early lactation and increase in C18:1 cis-9 in mid and late lactation, in response to the energy requirement). This double effect of pasture quality and animal lactation stage on the FA composition of cheese (particularly in sheep cheese) was recently reported by [48]. As aforementioned, it is likely to be a consequence of the typical seasonality of lambing of Sarda dairy sheep.

PC2 had larger negative loadings for rumenic acid, C4:0 and some OBCFA, in particular, anteisoC15:0, C15:0, and some isomer of C18:1, including vaccenic acid. Higher positive loadings of PC2 were observed for C18:0 and C10:0. This pattern suggested a better FA composition of sheep cheese, from a nutritional point of view, considering that FA with negative loadings for PC2 (including some OBCFA), and, therefore, mostly related to sheep cheese, have been associated to beneficial aspect on human health, that were the subject of this work. Due to its nutritional importance, the rumenic acid (most abundant and important CLA isomer) should be mentioned. This FA has been received particular attention due to the important beneficial healthy effects shown in vitro, animal and human as well [50,51]. The higher concentration in sheep milk (and cheese) compared to that of goat has previously been reported [45].

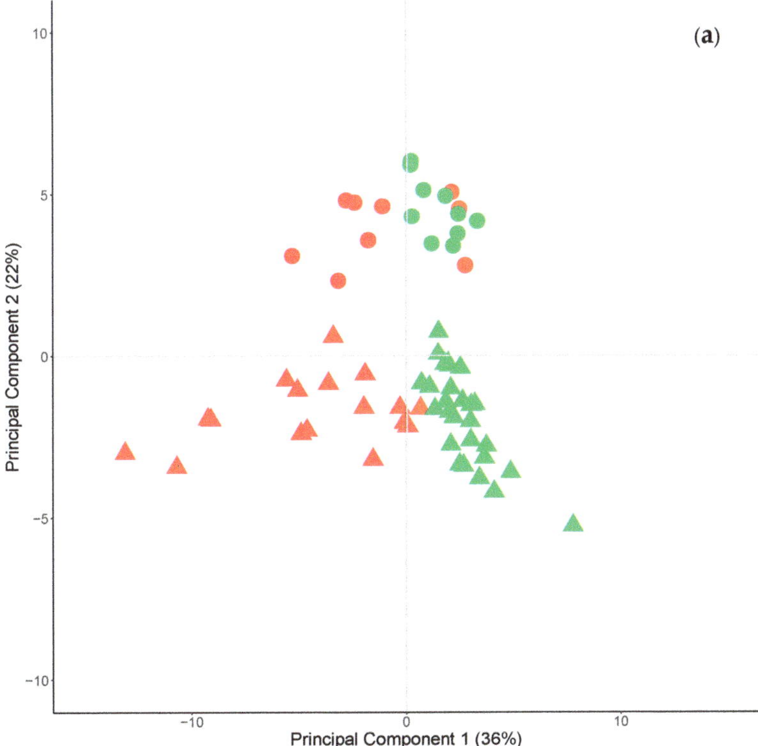

Figure 2. *Cont.*

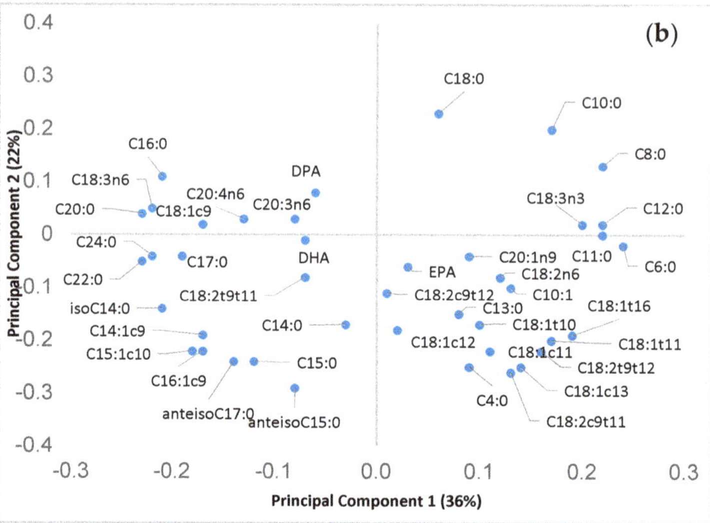

Figure 3. Scores (**a**) and loadings (**b**) plots of the two first principal components explaining 36 and 22% of the total variance, respectively. In Figure 3a, circle identifies goat cheese, triangle identifies sheep cheese; green symbols identified cheese produced in winter early spring, red symbols identified cheese produced in early spring–summer.

4. Conclusions

Results of the present survey showed that odd and branched chain fatty acids concentrations in the fat of fresh cheese is related to their content in both sheep and goats' milk. The seasonal evolution evidenced that OBCFA concentrations in milk fat increase with advancing lactation, probably due to variation in feeding technique typical of extensive system of Mediterranean area characterized by animal grazing on pasture. As lactation progress, the pasture availability and quality gets worse, leading to a reduction in the content of PUFA and protein and to an increase in fiber content. These results showed that cheese could be an important source of OBCFA, but the concentration could vary according to the lactation stage. The use of PCA on the detailed FA profile of cheese confirmed the higher nutritional quality of sheep cheese for beneficial FA including OBCFA compared to the goat one, and the importance of the period of sampling in the definition of the fatty acids profile.

Author Contributions: Individual contributions: A.N.: Conceptualization, Statistics and Writing. F.C.: Methodology and Analysis. A.C.: Software and PCA analysis. G.P.: Supervision and Writing. G.B.: Writing Original Draft, Preparation. All authors have read and agreed to the published version of the manuscript.

Funding: This research was funded by "Fondo di Ateneo per la ricerca 2019 of University of Sassari".

Institutional Review Board Statement: Not applicable.

Informed Consent Statement: Not applicable.

Data Availability Statement: The data presented in this study are available on request from the corresponding author.

Conflicts of Interest: The authors declare no conflict of interest.

References

1. Dagnachew, B.S.; Meuwissen, T.H.E.; Ådnøy, T. Genetic components of milk Fourier-transform infrared spectra used to predict breeding values for milk composition and quality traits in dairy goats. *J. Dairy Sci.* **2013**, *96*, 5933–5942. [CrossRef] [PubMed]
2. Maroteau, C.; Palhière, I.; Larroque, H.; Clément, V.; Ferrand, M.; Tosser-Klopp, G.; Rupp, R. Genetic parameter estimation for major milk fatty acids in Alpine and Saanen primiparous goats. *J. Dairy Sci.* **2014**, *97*, 3142–3155. [CrossRef] [PubMed]
3. Cesarani, A.; Gaspa, G.; Correddu, F.; Cellesi, M.; Dimauro, C.; Macciotta, N.P.P. Genomic selection of milk fatty acid composition in Sarda dairy sheep: Effect of different phenotypes and relationship matrices on heritability and breeding value accuracy. *J. Dairy Sci.* **2019**, *102*, 3189–3203. [CrossRef] [PubMed]
4. Ran-Ressler, R.R.; Bae, S.; Lawrence, P.; Wang, D.H.; Brenna, J.T. Branched-chain fatty acid content of foods and estimated intake in the USA. *Br. J. Nutr.* **2014**, *112*, 565–572. [CrossRef] [PubMed]
5. Ran-Ressler, R.R.; Khailova, L.; Arganbright, K.M.; Adkins-Rieck, C.K.; Jouni, Z.E.; Koren, O.; Ley, R.E.; Brenna, J.T.; Dvorak, B. Branched chain fatty acids reduce the incidence of necrotizing enterocolitis and alter gastrointestinal microbial ecology in a neonatal rat model. *PLoS ONE* **2011**, *6*, e29032. [CrossRef]
6. Vlaeminck, B.; Fievez, V.; Tamminga, S.; Dewhurst, R.J.; Van Vuuren, A.; De Brabander, D.; Demeyer, D. Milk odd-and branched-chain fatty acids in relation to the rumen fermentation pattern. *J. Dairy Sci.* **2006**, *89*, 3954–3964. [CrossRef]
7. Massart-Leën, A.M.; Peeters, G.; Vandeputte-Van Messom, G.; Roets, E.; Burvenich, C. Effects of valerate and isobutyrate on fatty acid secretion by the isolated perfused mammary gland of the lactating goat. *Reprod. Nutr. Dev.* **1986**, *26*, 801–814. [CrossRef]
8. Berthelot, V.; Bas, P.; Pottier, E.; Normand, J. The effect of maternal linseed supplementation and/or lamb linseed supplementation on muscle and subcutaneous adipose tissue fatty acid composition of indoor lambs. *Meat Sci.* **2012**, *90*, 548–557. [CrossRef]
9. Brevik, A.; Veierød, M.B.; Drevon, C.A.; Andersen, L.F. Evaluation of the odd fatty acids 15: 0 and 17: 0 in serum and adipose tissue as markers of intake of milk and dairy fat. *Eur. J. Clin. Nutr.* **2005**, *59*, 1417–1422. [CrossRef]
10. Albani, V.; Celis-Morales, C.; O'Donovan, C.B.; Walsh, M.C.; Woolhead, C.; Forster, H.; Fallaize, R.; Macready, A.L.; Marsaux, C.F.M.; Navas-Carretero, S.; et al. Within-person reproducibility and sensitivity to dietary change of C15: 0 and C17: 0 levels in dried blood spots: Data from the European Food4Me Study. *Mol. Nutr. Food. Res.* **2017**, *61*, 1700142. [CrossRef]
11. Khaw, K.T.; Friesen, M.D.; Riboli, E.; Luben, R.; Wareham, N. Plasma phospholipid fatty acid concentration and incident coronary heart disease in men and women: The EPIC-Norfolk prospective study. *PLoS Med.* **2012**, *9*, e1001255. [CrossRef] [PubMed]
12. Jenkins, B.; West, J.A.; Koulman, A. A review of odd-chain fatty acid metabolism and the role of pentadecanoic Acid (c15:0) and heptadecanoic Acid (c17:0) in health and disease. *Molecules* **2015**, *20*, 2425–2444. [CrossRef]
13. Warensjö, E.; Jansson, J.H.; Cederholm, T.; Boman, K.; Eliasson, M.; Hallmans, G.; Johansson, I.; Sjögren, P. Biomarkers of milk fat and the risk of myocardial infarction in men and women: A prospective, matched case-control study. *Am. J. Clin. Nutr.* **2010**, *92*, 194–202. [CrossRef] [PubMed]
14. Sun, Q.; Ma, J.; Campos, H.; Hu, F.B. Plasma and erythrocyte biomarkers of dairy fat intake and risk of ischemic heart disease. *Am. J. Clin. Nutr.* **2007**, *86*, 929–937. [CrossRef]
15. Krachler, B.; Norberg, M.; Eriksson, J.W.; Hallmans, G.; Johansson, I.; Vessby, B.; Weinehall, L.; Lindahl, B. Fatty acid profile of the erythrocyte membrane preceding development of Type 2 diabetes mellitus. *Nutr. Metab. Cardiovasc. Dis.* **2008**, *18*, 503–510. [CrossRef]
16. Hodge, A.M.; English, D.R.; O'Dea, K.; Sinclair, A.J.; Makrides, M.; Gibson, R.A.; Giles, G.G. Plasma phospholipid and dietary fatty acids as predictors of type 2 diabetes: Interpreting the role of linoleic acid. *Am. J. Clin. Nutr.* **2007**, *86*, 189–197. [CrossRef]
17. Forouhi, N.G.; Koulman, A.; Sharp, S.J.; Imamura, F.; Kröger, J.; Schulze, M.B.; Crowe, F.L.; Huerta, J.M.; Guevara, M.; Beulens, J.W.; et al. Differences in the prospective association between individual plasma phospholipid saturated fatty acids and incident type 2 diabetes: The EPIC-InterAct case-cohort study. *Lancet Diabetes Endocrinol.* **2014**, *2*, 810–818. [CrossRef]
18. Imamura, F.; Fretts, A.; Marklund, M.; Ardisson Korat, A.V.; Yang, W.S.; Lankinen, M.; Qureshi, W.; Helmer, C.; Chen, T.A.; Wong, K.; et al. Fatty acid biomarkers of dairy fat consumption and incidence of type 2 diabetes: A pooled analysis of prospective cohort studies. *PLoS Med.* **2018**, *15*, e1002670. [CrossRef]
19. Venn-Watson, S.; Lumpkin, R.; Dennis, E.A. Efficacy of dietary odd-chain saturated fatty acid pentadecanoic acid parallels broad associated health benefits in humans: Could it be essential? *Sci. Rep.* **2020**, *10*, 8161. [CrossRef] [PubMed]
20. Correddu, F.; Gaspa, G.; Pulina, G.; Nudda, A. Grape seed and linseed, alone and in combination, enhance unsaturated fatty acids in the milk of Sarda dairy sheep. *J. Dairy Sci.* **2016**, *99*, 1725–1735. [CrossRef]
21. Baumann, E.; Chouinard, P.Y.; Lebeuf, Y.; Rico, D.E.; Gervais, R. Effect of lipid supplementation on milk odd- and branched-chain fatty acids in dairy cows. *J. Dairy Sci.* **2016**, *99*, 6311–6323. [CrossRef] [PubMed]
22. Vazirigohar, M.; Dehghan-Banadaky, M.; Rezayazdi, K.; Nejati-Javaremi, A.; Mirzaei-Alamouti, H.; Patra, A.K. Short communication: Effects of diets containing supplemental fats on ruminal fermentation and milk odd- and branched-chain fatty acids in dairy cows. *J. Dairy Sci.* **2018**, *101*, 6133–6141. [CrossRef] [PubMed]

23. Zhang, Y.; Liu, K.; Hao, X.; Xin, H. The relationships between odd-and branched-chain fatty acids to ruminal fermentation parameters and bacterial populations with different dietary ratios of forage and concentrate. *J. Anim. Physiol. Anim. Nutr.* **2017**, *101*, 1103–1114. [CrossRef] [PubMed]
24. Correddu, F.; Nudda, A.; Battacone, G.; Boe, R.; Francesconi, A.H.D.; Pulina, G. Effects of grape seed supplementation, alone or associated with linseed, on ruminal metabolism in Sarda dairy sheep. *Anim. Feed. Sci. Technol.* **2015**, *199*, 61–72. [CrossRef]
25. Stoop, W.M.; Bovenhuis, H.; Heck, J.M.; van Arendonk, J.A. Effect of lactation stage and energy status on milk fat composition of Holstein-Friesian cows. *J. Dairy Sci.* **2009**, *92*, 1469–1478. [CrossRef]
26. De Souza, J.; Leskinen, H.; Lock, A.L.; Shingfield, K.J.; Huhtanen, P. Between-cow variation in milk fatty acids associated with methane production. *PLoS ONE* **2020**, *15*, e0235357. [CrossRef]
27. Craninx, M.; Steen, A.; Van Laar, H.; Van Nespen, T.; Martin-Tereso, J.; De Baets, B.; Fievez, V. Effect of lactation stage on the odd-and branched-chain milk fatty acids of dairy cattle under grazing and indoor conditions. *J. Dairy Sci.* **2008**, *91*, 2662–2677. [CrossRef]
28. Nudda, A.; McGuire, M.A.; Battacone, G.; Pulina, G. Seasonal variation in conjugated linoleic acid and vaccenic acid in milk fat of sheep and its transfer to cheese and ricotta. *J. Dairy Sci.* **2005**, *88*, 1311–1319. [CrossRef]
29. Santillo, A.; Caroprese, M.; Marino, R.; d'Angelo, F.; Sevi, A.; Albenzio, M. Fatty acid profile of milk and Cacioricotta cheese from Italian Simmental cows as affected by dietary flaxseed supplementation. *J. Dairy Sci.* **2016**, *99*, 2545–2551. [CrossRef]
30. IDF Standard 182:1999. *Milk Fat: Preparation of Fatty Acid Methyl Esters*; IDF: Brussels, Belgium, 1999.
31. Gómez-Cortés, P.; Rodríguez-Pino, V.; Juárez, M.; de la Fuente, M.A. Optimization of milk odd and branched-chain fatty acids analysis by gas chromatography using an extremely polar stationary phase. *Food Chem.* **2017**, *231*, 11–18. [CrossRef]
32. Vetter, W.; Gaul, S.; Thurnhofer, S.; Mayer, K. Stable carbon isotope ratios of methyl-branched fatty acids are different to those of straight-chain fatty acids in dairy products. *Anal. Bioanal. Chem.* **2007**, *389*, 597–604. [CrossRef]
33. Toral, P.G.; Hervás, G.; Della Badia, A.; Gervais, R.; Frutos, P. Effect of dietary lipids and other nutrients on milk odd- and branched-chain fatty acid composition in dairy ewes. *J. Dairy Sci.* **2020**, *103*, 11413–11423. [CrossRef] [PubMed]
34. Bainbridge, M.L.; Cersosimo, L.M.; Wright, A.D.; Kraft, J. Content and Composition of Branched-Chain Fatty Acids in Bovine Milk Are Affected by Lactation Stage and Breed of Dairy Cow. *PLoS ONE* **2016**, *11*, e0150386. [CrossRef]
35. Ran-Ressler, R.R.; Sim, D.; O'Donnell-Megaro, M.; Bauman, D.E.; Barbano, D.M.; Brenna, J.T. Branched chain fatty acid content of United States retail cow's milk and implications for dietary intake. *Lipids* **2011**, *46*, 569–576. [CrossRef] [PubMed]
36. Xin, H.; Xu, Y.; Chen, Y.; Chen, G.; Steele, M.A.; Guan, L.L. Short communication: Odd-chain and branched-chain fatty acid concentrations in bovine colostrum and transition milk and their stability under heating and freezing treatments. *J. Dairy Sci.* **2020**, *103*, 11483–11489. [CrossRef] [PubMed]
37. Fievez, V.; Vlaeminck, B.; Dhanoa, M.S.; Dewhurst, R.J. Use of principal component analysis to investigate the origin of heptadecenoic and conjugated linoleic acids in milk. *J. Dairy Sci.* **2003**, *86*, 4047–4053. [CrossRef]
38. Macciotta, N.P.P.; Cappio-Borlino, A.; Pulina, G. Analysis of environmental effects on test day milk yields of Sarda dairy ewes. *J. Dairy Sci.* **1999**, *82*, 2212–2217. [CrossRef]
39. Cabiddu, A.; Decandia, M.; Addis, M.; Piredda, G.; Pirisi, A.; Molle, G. Managing Mediterranean pastures in order to enhance the level of beneficial fatty acids in sheep milk. *Small Rumin. Res.* **2005**, *59*, 169–180. [CrossRef]
40. Chilliard, Y.; Ferlay, A.; Mansbridge, R.M.; Doreau, M. Ruminant milk fat plasticity: Nutritional control of saturated, polyunsaturated, trans and conjugated fatty acids. *Ann. Zootech.* **2000**, *49*, 181–205. [CrossRef]
41. Lourenço, M.; Ramos-Morales, E.; Wallace, R.J. The role of microbes in rumen lipolysis and biohydrogenation and their manipulation. *Animal* **2010**, *4*, 1008–1023. [CrossRef] [PubMed]
42. Enjalbert, F.; Combes, S.; Zened, A.; Meynadier, A. Rumen microbiota and dietary fat: A mutual shaping. *J. Appl. Microbiol.* **2017**, *123*, 782–797. [CrossRef] [PubMed]
43. Maia, M.R.; Chaudhary, L.C.; Figueres, L.; Wallace, R.J. Metabolism of polyunsaturated fatty acids and their toxicity to the microflora of the rumen. *Antonie Van Leeuwenhoek* **2007**, *91*, 303–314. [CrossRef] [PubMed]
44. Bas, P.; Archimède, H.; Rouzeau, A.; Sauvant, D. Fatty acid composition of mixed-rumen bacteria: Effect of concentration and type of forage. *J. Dairy Sci.* **2003**, *86*, 2940–2948. [CrossRef]
45. Nudda, A.; Cannas, A.; Correddu, F.; Atzori, A.S.; Lunesu, M.F.; Battacone, G.; Pulina, G. Sheep and Goats Respond Differently to Feeding Strategies Directed to Improve the Fatty Acid Profile of Milk Fat. *Animals* **2020**, *10*, 1290. [CrossRef]
46. Kadegowda, A.K.G.; Piperova, L.S.; Erdman, R.A. Principal component and multivariate analysis of milk long-chain fatty acid composition during diet-induced milk fat depression. *J. Dairy Sci.* **2008**, *91*, 749–759. [CrossRef]
47. Chung, I.M.; Kim, J.K.; Lee, K.J.; Son, N.Y.; An, M.J.; Lee, J.H.; An, Y.J.; Kim, S.H. Discrimination of organic milk by stable isotope ratio, vitamin E, and fatty acid profiling combined with multivariate analysis: A case study of monthly and seasonal variation in Korea for 2016–2017. *Food Chem.* **2018**, *261*, 112–123. [CrossRef]
48. Correddu, F.; Murgia, M.A.; Mangia, N.P.; Lunesu, M.F.; Cesarani, A.; Deiana, P.; Pulina, G.; Nudda, A. Effect of altitude of flock location, season of milk production and ripening time on the fatty acid profile of Pecorino Sardo cheese. *Int. J. Dairy Technol.* **2021**, *113*, 104895. [CrossRef]

49. Gómez-Cortés, P.; Cívico, A.; de la Fuente, M.A.; Núñez Sánchez, N.; Peña Blanco, F.; Martínez Marín, A.L. Effects of dietary concentrate composition and linseed oil supplementation on the milk fatty acid profile of goats. *Animal* **2018**, *12*, 2310–2317. [CrossRef]
50. Yang, B.; Chen, H.; Stanton, C.; Ross, R.P.; Zhang, H.; Chen, Y.Q.; Chen, W. Review of the roles of conjugated linoleic acid in health and disease. *J. Funct. Foods* **2015**, *15*, 314–325. [CrossRef]
51. Ferlay, A.; Bernard, L.; Meynadier, A.; Malpuech-Brugère, C. Production of trans and conjugated fatty acids in dairy ruminants and their putative effects on human health: A review. *Biochimie* **2017**, *141*, 107–120. [CrossRef]

Article

LC-QTOF/MS Untargeted Metabolomics of Sheep Milk under Cocoa Husks Enriched Diet

Cristina Manis [1], Paola Scano [1], Anna Nudda [2], Silvia Carta [2], Giuseppe Pulina [2] and Pierluigi Caboni [1,*]

[1] Dipartimento di Scienze della Vita e dell'Ambiente, Campus Universitario, 09042 Monserrato, Italy; cristina.manis@unica.it (C.M.); scano@unica.it (P.S.)

[2] Dipartimento di Agraria, Sezione di Scienze Zootecniche, Università degli Studi di Sassari, Viale Italia 39, 07100 Sassari, Italy; anudda@uniss.it (A.N.); scarta2@uniss.it (S.C.); gpulina@uniss.it (G.P.)

* Correspondence: caboni@unica.it; Tel.: +39-070-6758617

Abstract: The aim of this work was to evaluate, by an untargeted metabolomics approach, changes of milk metabolites induced by the replacement of soybean hulls with cocoa husks in the ewes' diet. Animals were fed with a soybean diet integrated with 50 or 100 g/d of cacao husks. Milk samples were analyzed by an ultra high performance liquid chromatograph coupled to a time of flight mass spectrometer (UHPLC-QTOF-MS) platform. Multivariate statistical analysis showed that the time of sampling profoundly affected metabolite levels, while differences between treatments were evident at the fourth week of sampling. Cocoa husks seem to induce level changes of milk metabolites implicated in the thyroid hormone metabolism and ubiquinol-10 biosynthesis.

Keywords: mass spectrometry; animals management; thyroid hormone metabolism; ubiquinol-10 biosynthesis

Citation: Manis, C.; Scano, P.; Nudda, A.; Carta, S.; Pulina, G.; Caboni, P. LC-QTOF/MS Untargeted Metabolomics of Sheep Milk under Cocoa Husks Enriched Diet. *Dairy* **2021**, *2*, 112–121. https://doi.org/10.3390/dairy2010011

Received: 15 December 2020
Accepted: 26 January 2021
Published: 22 February 2021

Publisher's Note: MDPI stays neutral with regard to jurisdictional claims in published maps and institutional affiliations.

1. Introduction

In the Mediterranean basin, where dairy sheep and goats breeding is widespread, the use of agricultural by-products in the diet of small ruminants is an ancient practice. Today the great availability of agro-industrial by-products produced worldwide opens the door to other products to be tested. Noteworthy, the use of co-feeds can contribute to reducing the impact associated with ruminant's management. Several positive aspects were observed when agro-industrial by-products (e.g., grape, olive, tomato, citrus pulp and myrtle residues) were included in the diet of small dairy ruminants. In particular, it has been reported that the diet implementation with these by-products has beneficial effects on ruminal metabolism, animal health, and quality of derived products [1–3].

Cocoa husks represent the part of cocoa pod left over and are the principal by-products derived from *Theobroma cacao* L., representing an important crop for many tropical developing countries. Cocoa husks are obtained after the removal of the cocoa beans, the principal commercial product of cocoa processing [3]. Husks represent 70–75% of the fruit weight [4] and, besides limited local use, are considered as waste [5]. Cocoa pod husks along with other industrial by-products such as cocoa bean shells, cocoa bean meal and cocoa germ can be used as feed especially in the developing countries.

Purine alkaloids, including theobromine, theophylline and caffeine, and other methylxanthines, play a substantial role in pharmacology and food chemistry. A limited number of plant species accumulates purine alkaloids, such as caffeine and theobromine, which are synthesized from xanthosine, a catabolite of purine nucleotides. The most widely distributed methylxanthine in the plant kingdom is the antioxidant caffeine, which accumulates in leaves and seeds of tea (*Camellia sinensis*), coffee (*Coffea arabica*) and a limited number of other species. Considerable amounts of theobromine are stored in the seeds of cacao (*Theobroma cacao*) [6]. Moreover, from a pharmacokinetic point of view, theobromine can be easily adsorbed, with a low protein binding affinity and distributed into body tissues with no specific tissue accumulation [7]. Despite the positive health benefits of consuming foods containing purine alkaloids [8] one of the limits in the use of cocoa husks as supplementary feed for animals is the fact that in certain amount, theobromine, can be toxic [9]. The EU Scientific Opinion of the Panel on Contaminants in the Food Chain reported that when exposed to theobromine dairy cows showed reduced milk yield, increase in fat levels, and adverse effects such as hyperexcitability, sweating, increase in respiration and heart rates [10]. Reported levels of theobromine in cocoa husk meal are in the range of 1.5–4.0 g per kg of feed material and UE regulations set for ruminants the maximum levels of theobromine in feed materials at 300 mg/kg. On the other side, cocoa husks on a dry matter basis (DM) are a good source of neutral detergent fiber (NFC 38–44%) and high variable concentrations of non-fiber carbohydrate (NFC 17.5–47% on DM), crude proteins (CP 2.1–9.1% on DM), and lipids (ethyl esters 0.6–4.7% on DM); phenolic compounds account for 4.6–6.9% of gallic acid/100 g DM. However, the high content of acid detergent lignin (19.6% on DM) derived from cocoa pericarp affects the digestibility of this by-product, limiting its further use in animal feeding [11].

Metabolomics is a largely used branch of omics sciences used to explore the composition, relative levels, dynamics, and interactions amongst metabolites in an organism or biological system in response to various stimuli, diet or treatments. In the field of dairy, metabolomics has been used for example in the authentication of beef production systems [12], in the milk to investigate metabolite differences of healthy, subclinical, and clinical mastitis cows [13] or sheep milk grazed on different grazing systems [14], while in cheese samples used for discriminate cheese produced from raw or thermized milk [15].

To our knowledge, experimental data on theobromine carry over in milk are not available, as well as the metabolic effects of cocoa husks diet supplementation. In a previous work, milk and blood parameters of ewes supplemented with different doses of cocoa husks were evaluated [5]. In this experiment DMI and milk yield were not affected; however, milk protein content increased with cocoa husk supplementation. Furthermore, the cocoa husks supplementation lowered milk somatic cell count (SCC), suggesting a beneficial role of these by-products on mammary health status. Both systemic and local health conditions were good as evidenced by hematological parameters and electrophoretic profile of serum protein fractions [5]. Both systemic and local health conditions were good as evidenced by hematological parameters and electrophoretic profile of serum protein fractions [5]. Taking into consideration these results, we decided to investigate the potential beneficial nutritional outcomes by exploring, for the first time, milk metabolite profile changes of healthy ewes fed with a diet supplemented with 100 g/d per head soybean hulls that were replaced with 50 and 100 g/d per head of cocoa husks, respectively. In order to pursue these objectives, we used a liquid chromatography mass spectrometry metabolomics platform and multivariate statistical analysis.

2. Materials and Methods

The experiment was approved by the Ethics Committee of the University of Sassari (no. 54584/2018).

Animals and diets. As previously described [5], twenty-four Sarda dairy ewes in mid lactation stage (120 days in milking), at third parity, with an average body weight of 42.5 kg were divided into three homogenous groups. Each experimental group received a total mixed ration (2.58 kg/d per head) as the basal diet and received (a) a supplement of 100 g/d per head of soybean hulls (CH0 group); (b) half of the soybean hulls supplement was replaced with 50 g/d per head of cocoa husks (CH50 group); and (c) soybean hulls was totally replaced with 100 g/d per head of cocoa husks (CH100 group). Theobromine concentration was 130 ± 12 and 253 ± 27 mg/kg of DM for the CH50 and CH100 diets, respectively. The CH was administered individually during milking, mixed with beet pulp to increase palatability. The experiment lasted 8 weeks (from May to June 2018), with the last 4 weeks of sampling collection. Individual milk samples were collected weekly at morning milking.

Chemicals. Sigma Aldrich (Milan, Italy) reagents were used for the analysis. Bi-distilled water was obtained with a MilliQ purification system (Millipore, Milan, Italy).

Sample preparation for UHPLC-QTOF/MS analysis. Individual milk samples from the morning milking were stored into sterile plastic Falcon tubes at −80 °C before analysis. Prior to UHPLC-QTOF/MS analysis, the 107 individual sheep milk samples were thawed and thoroughly vortex mixed. A total of 150 μL of milk samples were transferred to Eppendorf tube containing 10 uL of the internal mixture of standards (Splash, Lipidomics, Sigma Aldrich, Milan, Italy) and added with 525 μL of methanol and 525 μL of MTBE. Samples were then mixed by vortexing for 1 min. Next, samples were centrifuged at 4000 rpm for 15 min. Subsequently, a second round of centrifuge was performed at 12000 rpm for 5 min. Before transferring to autosampler vials, the supernatant was filtered through a 0.22 μm MS nylon syringe filter.

UHPLC-QTOF/MS analysis. After extraction with MTBE/methanol, the supernatant of milk samples was analyzed with a 6560 Q-TOF (Agilent Technologies, Palo Alto, CA, USA) coupled with an Agilent 1290 Infinity II LC system. An aliquot of 1.0 μL from each sample was injected in a Kinetex 5 μm EVO C18 100 A, 150 mm × 2.1 μm column (Agilent Technologies, Palo Alto, CA, USA). The column was maintained at 50 °C at a flow rate of 0.5 mL/min. The mobile phases consisted of (A) methanol:water (90:10, *v/v*) with ammonium acetate (10 mM) and (B) acetonitrile:methanol:2-propanol (20:30:50, *v/v*) with ammonium acetate (10 mM). The chromatographic separation was obtained using the following gradient—0 min 70% B kept for 1 min; 1–3.5 min 86% B; 3.5–10 min 86% B; 10.1–17 min 100% B; 17.1–19 min 70% B. The analytical setup used was equipped with an Agilent jet stream technology source which was operated in both positive and negative ion modes with the following parameters: gas temperature,200 °C; gas flow (nitrogen) 10 L/min; nebulizer gas (nitrogen), 50 psig; sheath gas temperature, 300 °C; sheath gas flow, 12 L/min; capillary voltage 3500 V for positive and 3000 V for negative; nozzle voltage 0 V; fragmentor 150 V; skimmer 65 V, octapole RF 7550 V; mass range, 40–1700 *m/z*; capillary voltage, 3.5 kV; collision energy 20 eV in positive and 25 eV in negative mode. An Agilent MassHunter software was used for instrument control (revision B.09.00).

Multivariate statistical data analysis (MVA). The QTOF-MS data were submitted to the web platform XCMS [16]. The workflow allows feature detection, retention time correction, alignment, annotation, statistical analysis, and data visualization. Retention time, *m/z* values, and intensities of each feature obtained from XCMS were then submitted to multivariate statistical analysis (MVA), as implemented in SIMCA-P+ software (version 14.1, Umetrics, Umeå, Sweden). Prior to MVA, QTOF-MS features were mean centered and scaled to unit variance column-wise. Principal component analysis (PCA) was performed to investigate sample distributions, deviating features and prevailing trends. This technique reduced the dimensionality of data set retaining most of the variance. Results are shown as scatter plots of scores and loadings in the first principal components, reporting sample and variable displacement in the hyperplane, respectively. The partial least squares-discriminant analysis (PLS-DA) and its orthogonal variant (OPLS-DA) were performed for classification of samples and identification of the most discriminant variables. The quality of the models was evaluated based on the cumulative parameters R^2Y and Q^2Y, the latter estimated by cross validation, and tested for overfitting using a y-table permutation test ($n = 400$). The variable importance in projection (VIP) score summarizes the contribution of each variable to the model. The VIP scores in the predictive component were analyzed and only those metabolites having VIP values > 1 were considered as discriminant between the classes [17].

3. Results and Discussion

This untargeted metabolomics approach on dairy ewes was conceived to provide a molecular and biological insights on milk metabolite changes induced by a diet integrated with 50 or 100 g/d of cacao husks. Moreover, we designed this cost-efficient metabolomics study to measure predictive metabolite biomarkers to be used in exploring feed suitability and efficiency.

For these purposes, a total of 107 sheep milk samples were analyzed with a UHPLC-QTOF/MS platform, (Figure S1 reported the LC-MS chromatograms). After analysis, raw mass spectrometry data acquired were then uploaded to the XCMS platform, which generated an average of 7700 and 3300 features for the positive and negative modes, respectively. The XCMS outputs were submitted to multivariate statistical analysis using the software SIMCA.

A PCA was performed to explore similarities of milk sample metabolic profiles of the three diets over the experimental sampling period. Results indicated that metabolic similarities of milk samples were mainly due to the time of sampling, as clearly visible in the score plots shown in Figure 1 for the positive and negative ionization modes.

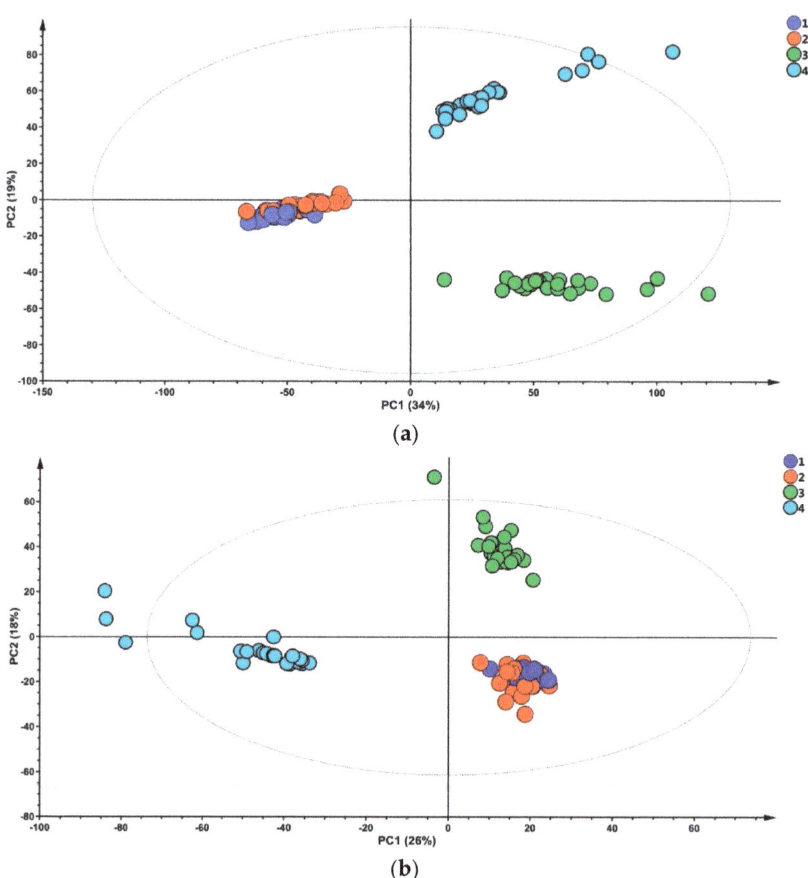

Figure 1. Principal component analysis (PCA) score plots of milk samples. Samples are colored by week of treatment (first, blue; second, red; third, green; and fourth, light blue circles). Metabolite data collected in the liquid chromatography-mass spectrometry (LCMS) (**a**) positive and (**b**) negative ionization modes.

Information collected for the two ionization modes yielded similar results (i.e., samples collected in the two first weeks of treatment clustered together as a result of similar metabolite profile, while milk samples from the third and the last week of experiment can be easily differentiated along the second principal component (PC2) in the positive and negative ionization modes). Interestingly, the same pattern was observed for milk yield, as shown in Figure S2, whereby the first two weeks gave comparable average values for all the group samples, followed by a drastic milk yield drop after the second week of sampling (from the 13 to 26 of June). A decrease of milk yield and lactose content and an increase in fat and protein content, due to a milk concentration effect, during the late-spring/summer seasons were already reported as a natural evolution of the lactation curve in dairy Sarda sheep [18]. Probably these milk qualitative and quantitative changes due to lactation seasonality have caused an overall modification of the metabolic profiles captured by the PCA.

Attempts to differentiate samples based on the diets (CH0, CH50, and CH100) were carried out by performing three-classes PLS-DA for each time point for positive and negative modes. The multivariate statistical analysis failed to discriminate samples for the first three weeks of sampling (data not shown) whereas at the fourth week of treatment, samples were properly classified, with high statistical confidence (Figure S3), indicating that the effects of the cocoa husks supplementation on milk metabolic profile were evident only in the last time point. Subsequently, with the purpose of finding discriminant metabolites, pairwise OPLS-DA were carried out comparing the three groups of milk samples collected in the last week of treatment (Figure 2). In Tables 1 and 2 we reported the list of discriminant metabolites, their biological significance, and regulation in sheep milk samples.

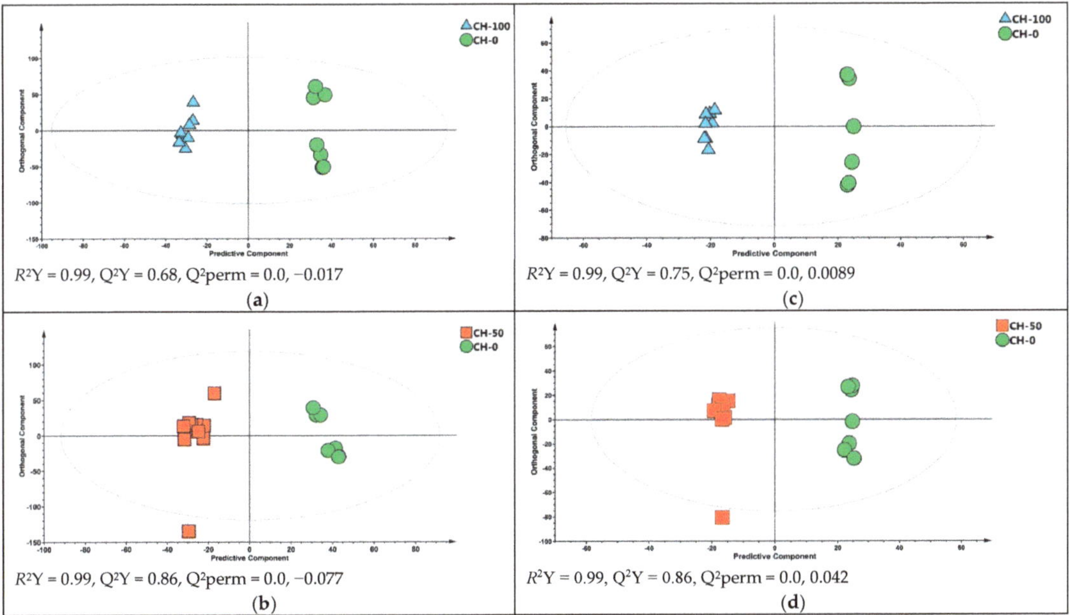

$R^2Y = 0.99$, $Q^2Y = 0.68$, Q^2perm = 0.0, −0.017

(a)

$R^2Y = 0.99$, $Q^2Y = 0.75$, Q^2perm = 0.0, 0.0089

(c)

$R^2Y = 0.99$, $Q^2Y = 0.86$, Q^2perm = 0.0, −0.077

(b)

$R^2Y = 0.99$, $Q^2Y = 0.86$, Q^2perm = 0.0, 0.042

(d)

Figure 2. Score plots of pairwise orthogonal partial least squares-discriminant analysis (OPLS-DA) of milk samples data in liquid chromatography-mass spectrometry positive ((**a**) CH0 vs. CH100 and (**b**) CH50 vs. CH0), and negative ((**c**) CH0 vs. CH100 and (**d**) CH50 vs. CH0) ionization modes. CH0 = green circles; CH50 = red boxes; CH100 = light blue triangles.

Table 1. Orthogonal partial least squares-discriminant analysis (OPLS-DA) main discriminant metabolites as variable importance for prediction (VIP > 1) detected in positive ionization modes.

RT (min)	*m/z*	Adduct	Positive Ionization Mode Metabolite	VIP	Regulation	Pathway Involved
1.14	279.1603	$[M+H]^+$	α-carboxyethyl-hydroxychroman	2.21	down	Vitamin E metabolism
6.57	707.5197	$[M+K]^+$	2-methoxy-6-(all-*trans*-octaprenyl)phenol	1.93	down	Ubiquinol-10 biosynthesis
5.77	717.5652	$[M-H_2O+H]^+$	1,2-dipalmitoyl-phosphatidylcholine	1.81	down	
1.26	631.4714	$[M-H_2O+H]^+$	dipalmitoyl phosphatidate	1.80	down	
0.64	336.0713	$[M+H]^+$	β-nicotinamide D-ribonucleotide	1.55	up	NADH synthesis
0.61	145.0504	$[M-H_2O+H]^+$	1,5-anhydro-D-fructose	1.43	up	
0.64	158.0415	$[M+2H]^{2+}$	geranyl diphosphate	1.40	up	
0.63	118.0876	$[M-NH_3+H]^+$	L-ornithine	1.26	up	Urea cycle
0.66	321.0601	$[M-H_2O+H]^+$	5-amino-1-(5-phospho-D-ribosyl)imidazole-4-carboxamide	1.15	up	Purine metabolism
8.69	817.6885	$[M-H_2O+H]+$	6-methoxy-3-methyl-2-all-*trans*-decaprenyl-1,4-benzoquinol	1.07	down	Ubiquinol-10 biosynthesis

Table 2. Orthogonal partial least squares-discriminant analysis (OPLS-DA) main discriminant metabolites (VIP>1) detected in negative ionization modes.

RT (min)	*m/z*	Adduct	Negative Ionization Mode Metabolite	VIP	Regulation	
0.62	179.0572	$[M-H]^-$	theobromine	2.43	up	
0.64	211.0842	$[M-H]^-$	phenylalanine	2.34	up	
6.17	710.6315	$[M-2H_2O-H]^-$	tetraiodothyroacetate	1.75	down	Thyroid hormone synthesis
6.48	833.6484	$[M-H]^-$	3,4-dihydroxy-5-all-*trans*-decaprenylbenzoate	1.61	down	Ubiquinol-10 biosynthesis
6.15	812.6624	$[M+Cl]^-$	L-thyroxine	1.60	down	Thyroid hormone synthesis
11.38	369.3014	$[M+CH_3COO]^-$	phytenate	1.58	down	Phytol degradation
0.72	815.0952	$[M+CH_3COO]^-$	5′,5‴-diadenosine triphosphate	1.49	up	Purine metabolism
0.65	259.0235	$[M-H]^-$	hexose 6-phosphate	1.40	up	
0.64	478.9864	$[M+Na-2H]^-$	8-oxo-GDP	1.36	up	
0.61	267.0740	$[M-H]^-$	inosine	1.20	up	Purine metabolism
7.33	311.2968	$[M-H]^-$	arachidic acid	1.08	up	
11.41	489.3955	$[M+CH_3COO]^-$	α-tocopherol	1.03	down	
0.66	142.0214	$[M-2H_2O-H]^-$	L-cysteinyl-glycine	1.00	up	
1.58	435.2918	$[M+CH_3COO]^-$	all-*trans*/10′-apo-β-carotenal	1.00	up	Carotenoid cleavage

From the analysis of discriminant metabolites, as summarized in Tables 1 and 2, we found that the addition of cocoa husks in the dairy lactating sheep diet affected levels of different metabolites which are involved in different metabolic pathways, including ubiquinol synthesis and thyroid hormone metabolism. In particular, we found that in milk samples of lactating ewes the thyroid hormone L-thyroxine (T4) and tetraiodothyroacetate, or tetrac (T4A), were down regulated for both CH50 and CH100 groups, thus suggesting a modification of the thyroid hormone metabolism. Levels of T4 and T4A are shown in Figure 3, indicating that, in the first two weeks of treatment, group samples showed similar average values. Levels where higher in the third week and in the last week CH50 and CH100 groups showed lower values when compared to CH0.

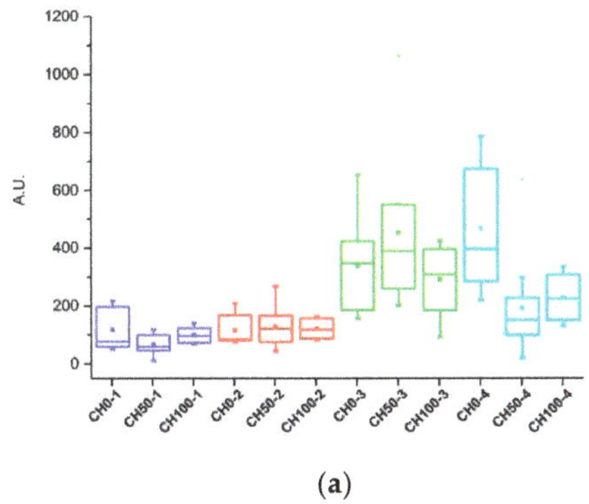 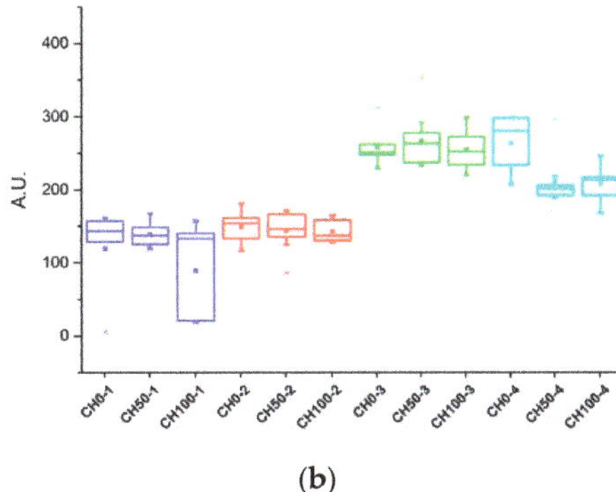

(a)   (b)

Figure 3. Box plots of (**a**) thyroxine (T4) and (**b**) tetraiodothyroacetate (T4A) levels in milk samples for the CH0, CH50, and CH100 groups in the last four weeks of treatments (week 1 = blue, 2 = red, 3 = green, 4 = cyan).

The thyroid hormones T4 and 3,5,3′-triiodo-L-thyronine (T3) are regulated at many metabolic levels. In addition to deiodination (thyroid hormone metabolism I pathway), other routes of thyroid hormone metabolism have been described and metabolically active metabolites have been identified. In the thyroid hormone metabolism pathway (via degradation), small amounts of T4 may also be oxidatively deaminated at the alanine side chain to yield T4A. The oxidative deamination due the kidney L-amino acid oxidase seems the most probable explanation [19,20]. The acetic acid analogues of thyroid hormones are formed in the liver and other peripheral tissue by a two-stage process involving transamination followed by oxidative decarboxylation [21]. T4A may be conjugated to form ether glucuronides or ester glucuronides of these metabolites may form by conjugation of the phenolic hydroxyl group, or carboxyl group respectively [20].

Thyroid hormones in milk should derive from circulating blood, then undergoing both desiodative and non-desiodative metabolism. A reduction in milk T4 could; therefore, be derived from a reduction in circulating T4 levels or from the reduction in T4 level available at the mammary gland, due to the effect of interferents on the transporters of thyroid hormones, or, finally, by increased desiodation of T4. The reduction of milk T4A level seems the logical direct consequence of the reduction of T4, deriving the T4A directly from T4. Thyroid hormones play a relatively important role in lactational processes. In small ruminants and, in particular, in free grazing animals which are subjected to seasonality, endogenous (breed, age, gender, physiological state) and environmental factors (climate, season, with a primary role of nutrition) are able to impact on thyroid activity and hormone blood levels [22]. During heat stress, an increase of blood T4 concentration and a decrease of milk production were reported in sheep [23,24]. This trend was evident also in our milk samples, which were collected in June, where, during the experimental period, in the CH0 group the milk T4 levels inversely follow the milk production trend (Figures S2 and S4). Differently from CH0, in the last week of sampling, milk T4 levels dropped in the CH50 and CH100 groups (Figure S4), thus indicating a goitrogenic effect of cocoa husks on milk T4. In the rat, Wolff and Varrone (1969) reported that methyl-xanthines (i.e., theobromine, theophylline, and caffeine) showed a goitrogenic effect and are; therefore, to be considered weak anti-thyroid agents if present in the diet at concentrations varying between 5–20 microM [25]. Further studies will be carried out to better evaluate this effect in sheep, measuring the volume of the thyroid glands and the concentration of T4 and TSH in ewes' blood after consumption of cocoa husks. At the milk level, which only contains traces of iodothyronines, which; therefore, do not represent a significant nutrient of the milk itself, as opposed to iodine, any reduction in concentration should not be particularly relevant for its nutritional value.

Notably, phenylalanine was found up-regulated in CH50 and CH100 milk samples. Considering that phenylalanine is the precursor of tyrosine and the thyroid hormone T4 is a tetraiodinated derivative of tyrosine [22], this fact confirms a dysregulation of thyroid hormones metabolism.

The discriminant metabolites: 2-methoxy-6-(all-trans-octaprenyl)phenol, 6-methoxy-3-methyl-2-all-trans-decaprenyl-1,4-benzoquinol, and 3,4-dihydroxy-5-all-trans-decaprenyl benzoate, involved in the ubiquinol-10 biosynthesis, were found down regulated in milk samples of the CH50 and CH100 groups with respect to CH0 (Table 1). Coenzyme Q (CoQ10), also known as ubiquinone, is an essential component of the mitochondrial respiratory chain and participates in aerobic cell respiration. The reduced ubiquinol form of CoQ10 is a chain-breaking antioxidant, decreasing oxidative damage caused by lipid peroxidation within mitochondria. The cocoa husks feeding regime, showed markedly lower levels of intermediate metabolites of the CoQ10 pathway compared to the controls. To the best of our knowledge, no studies of theobromine effects on CoQ10 pathway have been widely reported, but interference of phenolic compounds has been found in liver of rats [26]. Specifically, the authors found that pentagalloylglucose, which is a component of tannic acid, was able to interfere with the CoQ10 pathway: At low concentration inhibiting the electron transport system, while at high concentration impairing the structural integrity of the mitochondrial membrane. Consistently, CoQ10 was found to affect the pharmacokinetic parameters of theophylline [27], an alkaloid of the xanthine family.

Remarkably, in the negative ionization mode we found different discriminant metabolites involved in the purine metabolism (Table 1). Inosine is a purine nucleoside formed by deamination of adenosine and, in the context of purine metabolism, is further transformed to hypoxanthine by the enzyme phosphate alpha-D-ribosyltransferase and then to xanthine and to uric acid. The fact that inosine was found upregulated in milk samples of sheep consuming cocoa husks may be associated to the inhibition of the enzyme phosphate alpha-D-ribosyltransferase, in order to reduce the increase of xanthines and uric acid derived from the metabolism of theobromine.

Our experimental findings are in accordance with scientific opinion of EFSA [10] on theobromine as an undesired substance in animal feed. In the EFSA report, theobromine was reported as showing moderate acute toxicity on dog and rodents, while in horses, which are particularly susceptible to theobromine, the liver and thyroid were affected, and pigs showed growth retardation, diarrhea and lethargy. Particular attention should be paid on the levels of theobromine and other alkaloids present in the industrial by-products in the preparing of the dairy sheep's diet.

4. Conclusions

In this paper we show evidence, with an untargeted metabolomics approach, that sheep milk metabolite profiles are profoundly altered on the basis of an eight-week diet containing 50 and 100 g/d of cocoa husks. In particular these effects were more pronounced in the eighth week, at the end of the treatment. Moreover, this approach allowed us to highlight changes in multiple metabolic pathways. Importantly, the thyroid hormone L-thyroxine (T4) and tetraiodothyroacetate (T4A) were found down regulated for both CH50 and CH100 groups, thus suggesting a modification of the thyroid hormone metabolism. Furthermore, different ubiquinol-10 biosynthesis metabolites were found down regulated in milk samples of the CH50 and CH100. Biochemical mechanisms related to these changes remain unclear, while thyroid hormones levels should be further investigated in the ewes' blood. In conclusion, taking into account these results, particular attention should be paid when using cocoa by-products in small ruminant feed management.

Supplementary Materials: The following are available online at https://www.mdpi.com/2624-862X/2/1/11/s1: Figure S1. Milk sample overlaid chromatograms of CH0, CH50, and CH100 groups acquired in positive (a) and negative (b) ionization mode. Figure S2. Temporal evolution of milk yield in the control (CH0) and in the two cocoa husks supplemented groups (CH50 and CH100). Figure S3. Partial least square discriminant analysis (PLS-DA) score plots of milk samples of the experimental groups (CH0, CH50, and CH100) collected in the last week of treatment. Lipid data collected in the LCMS positive (top, R^2Y = 0.62, Q^2Y = 0.27) and negative (bottom, R^2Y = 0.98, Q^2Y = 0.63) ionization mode. Figure S4. Temporal evolution of thyroxine (T4) milk levels in the control (CH0) and in the two cocoa husks supplemented groups (CH50 and CH100).

Author Contributions: Sample analysis, statistical analysis, wrote original draft, C.M. and P.C.; statistical analysis, wrote original draft, P.S.; conceptualization, wrote original draft, A.N. and G.P.; animal experiment and milk sample collection, S.C. All authors have read and agreed to the published version of the manuscript.

Funding: The research was partly funded by Mignini-Petrini company.

Institutional Review Board Statement: The study was conducted according to the guidelines of the Declaration of Helsinki, and approved by the Ethics Committee of the University of Sassari (no. 54584/2018).

Informed Consent Statement: Not applicable.

Data Availability Statement: The data presented in this study are available on request from the corresponding author.

Acknowledgments: The authors are grateful to Stefano Mariotti and Elio Acquas for helpful discussions.

Conflicts of Interest: The authors declare that the sponsor had no role in the study design and interpretation of the results.

References

1. Mirzaei-Aghsaghali, A.; Maheri-Sis, N. Nutritive value of some agro-industrial by-products for ruminants—A review. *World J. Zool.* **2008**, *3*, 40–46.
2. Bampidis, V.A.; Robinson, P.H. Citrus by-products as ruminant feeds: A review. *Anim. Feed Sci. Technol.* **2006**, *128*, 175–217. [CrossRef]
3. Yusof, F.; Khanahmadi, S.; Amid, A.; Mahmod, S.S. Cocoa pod husk, a new source of hydrolase enzymes for preparation of cross-linked enzyme aggregate. *Springerplus* **2016**, *5*, 1–18. [CrossRef]
4. Cruz, G. Production of Activated Carbon from Cocoa (Theobroma cacao) Pod Husk. *J. Civ. Environ. Eng.* **2012**, *02*. [CrossRef]
5. Carta, S.; Nudda, A.; Cappai, G. Cocoa husks replace effectively soybean hulls in dairy sheep diet: Effects on milk production traits and metabolic profile. *J. Dairy Sci.* **2019**, 3–29.
6. Sato, F. *Plant Alkaloid Engineering*, 3rd ed.; Elsevier Inc.: Amsterdam, The Netherlands, 2019; ISBN 9780124095472.
7. Lelo, A.; Birkett, D.; Robson, R.; Miners, J. Comparative pharmacokinetics of caffeine and its primary demethylated metabolites paraxanthine, theobromine and theophylline in man. *Br. J. Clin. Pharmacol.* **1986**, *22*, 177–182. [CrossRef] [PubMed]
8. Martínez-Pinilla, E.; Oñatibia-Astibia, A.; Franco, R. The relevance of theobromine for the beneficial effects of cocoa consumption. *Front. Pharmacol.* **2015**, *6*, 1–5. [CrossRef] [PubMed]
9. Adamafio, N.A. Theobromine Toxicity and Remediation of Cocoa By-product: An Overview. *J. Biol. Sci.* **2013**, *13*, 570–576. [CrossRef]
10. Alexander, J.; Benford, D.; Cockburn, A.; Cravedi, J.; Dogliotti, E.; Domenico, A.D.; Férnandez-cruz, M.L.; Fürst, P.; Fink-gremmels, J.; Galli, C.L.; et al. Theobromine as undesirable substances in animal feed—Scientific opinion of the Panel on Contaminants in the Food Chain. *EFSA J.* **2008**, *6*, 1–66.
11. Campos-Vega, R.; Nieto-Figueroa, K.H.; Oomah, B.D. Cocoa (Theobroma cacao L.) pod husk: Renewable source of bioactive compounds. *Trends Food Sci. Technol.* **2018**, *81*, 172–184. [CrossRef]
12. Osorio, M.T.; Moloney, A.P.; Brennan, L.; Monahan, F.J. Authentication of beef production systems using a metabolomic-based approach. *Animal* **2012**, *6*, 167–172. [CrossRef]
13. Xi, X.; Kwok, L.Y.; Wang, Y.; Ma, C.; Mi, Z.; Zhang, H. Ultra-performance liquid chromatography-quadrupole-time of flight mass spectrometry MSE-based untargeted milk metabolomics in dairy cows with subclinical or clinical mastitis. *J. Dairy Sci.* **2017**, *100*, 4884–4896. [CrossRef]
14. Scano, P.; Carta, P.; Ibba, I.; Manis, C.; Caboni, P. An Untargeted Metabolomic Comparison of Milk Composition from Sheep Kept Under Different Grazing Systems. *Dairy* **2020**, *1*, 30–41. [CrossRef]
15. Caboni, P.; Maxia, D.; Scano, P.; Addis, M.; Dedola, A.; Pes, M.; Murgia, A.; Casula, M.; Profumo, A.; Pirisi, A. A gas chromatography-mass spectrometry untargeted metabolomics approach to discriminate Fiore Sardo cheese produced from raw or thermized ovine milk. *J. Dairy Sci.* **2019**, *102*, 5005–5018. [CrossRef] [PubMed]
16. Tautenhahn, R.; Patti, G.J.; Rinehart, D.; Siuzdak, G. XCMS online: A web-based platform to process untargeted metabolomic data. *Anal. Chem.* **2012**, *84*, 5035–5039. [CrossRef] [PubMed]

17. Eriksson, L.; Johansson, E.; Antti, H.; Holmes, E. Multi- and Megavariate Data Analysis. In *Metabonomics in Toxicity Assessment*; CRC Press: Boca Raton, FL, USA, 2005; pp. 263–336.
18. Macciotta, N.P.P.; Cappio-Borlino, A.; Pulina, G. Analysis of environmental effects on test day milk yields of Sarda dairy ewes. *J. Dairy Sci.* **1999**, *82*, 2212–2217. [CrossRef]
19. Wu, S.Y.; Green, W.L.; Huang, W.S.; Hays, M.T.; Chopra, I.J. Alternate pathways of thyroid hormone metabolism. *Thyroid* **2005**, *15*, 943–958. [CrossRef] [PubMed]
20. Moreno, M.; Kaptein, E.; Goglia, F.; Theo, T.J. Vesser Rapid Glucuronidation of Tri- and Tetraiodothyroacetic Acid to Ester Glucuronides in Human Liver and to Ether Glucuronides in Rat Liver. *Endocrinology* **1994**, *135*, 1004–1009. [CrossRef]
21. Crossley, D.N.; Ramsden, D.B. Serum tetraiodothyroacetate (t,a) levels in normal healthy euthyroid individuals determined by gas chromatography-mass fragmentography (gc-mf). *Clin. Chim. Acta* **1978**, *94*, 267–272. [CrossRef]
22. Todini, L. Thyroid hormones in small ruminants: Effects of endogenous, environmental and nutritional factors. *Animal* **2007**, *1*, 997–1008. [CrossRef]
23. Valtorta, S.; Hahn, L.; Johnson, H.D. Effect of High Ambient Temperature (35°), and Feed Intake on Plasma T_4 Levels in Sheep. *Proc. Soc. Exp. Biology Med.* **1982**, *169*, 260–265. [CrossRef] [PubMed]
24. Silanikove, N. Effects of heat stress on the welfare of extensively managed domestic ruminants. *Livest. Prod. Sci.* **2000**, *67*, 1–18. [CrossRef]
25. Wolff, J.; Varrone, S. The methyl xanthines—A new class of goitrogens. *Endocrinology* **1969**, *85*, 410–414. [CrossRef] [PubMed]
26. Adachi, H.; Konishi, K.; Horikoschi, I. The effects of 1,2,3,4,6-penta-O-galloyl-beta-D-glucose on rat liver mitochondrial respiration. *Chem. Pharm. Bull.* **2002**, *57*, 364–370.
27. Baskaran, R.; Shanmugam, S.; Nagayya-Sriraman, S.; Kim, J.H.; Jeong, T.C.; Yong, C.S.; Choi, H.G.; Yoo, B.K. The effect of coenzyme Q10 on the pharmacokinetic parameters of theophylline. *Arch. Pharm. Res.* **2008**, *31*, 938–944. [CrossRef] [PubMed]

Article

Quality Control in Fiore Sardo PDO Cheese: Detection of Heat Treatment Application and Production Chain by MRI Relaxometry and Image Analysis

Roberto Anedda *, Riccardo Melis and Elena Curti

Porto Conte Ricerche s.r.l., S.P. 55 Porto Conte-Capo Caccia, Km 8.400 Loc. Tramariglio, 07041 Alghero, SS, Italy; melis@portocontericerche.it (R.M.); curti@portocontericerche.it (E.C.)
* Correspondence: anedda@portocontericerche.it; Tel.: +39-079-998-578

Abstract: Fiore Sardo (FS), a traditional Italian cheese, is present in the market as a heterogeneous variety of products. The use of heat-treated (HT) milk is forbidden by the official production protocol, but no official analytical method able to detect heat application is yet available. Here, a combined magnetic resonance imaging (MRI) relaxometry and image analysis approach to recognize FS made from raw milk is presented. Artisanal FS cheeses were produced from raw milk (RC) by five shepherds in accordance with the official protocol. They were compared to HT-milk counterparts (HTC). Additionally, industrially manufactured commercial FS cheeses (I) were also purchased and compared to RC and HTC. Relaxometry data of FS indicated the presence of two water populations; the ratio of characteristic relaxation time constant T_2 and area fraction (Score, $) of the fastest relaxing population was used to compare RC, HTC and I samples. RC from HTC were successfully discriminated, the latter exhibiting lower $ (enhanced protein hydration). I cheeses exhibited the lowest $ values, sometimes comparable to HTC. Since visual appearance of RC and HTC is appreciably different, an image analysis deep learning approach using MRI and photographic pictures was adopted to discriminate the two productions, with promising percentages (>93%).

Keywords: magnetic resonance imaging (MRI); image analysis; Fiore Sardo; cheese; microstructure; dairy chemistry; thermised milk; raw milk; protected designation of origin

Citation: Anedda, R.; Melis, R.; Curti, E. Quality Control in Fiore Sardo PDO Cheese: Detection of Heat Treatment Application and Production Chain by MRI Relaxometry and Image Analysis. *Dairy* **2021**, *2*, 270–287. https://doi.org/10.3390/dairy2020023

Academic Editors: Pierluigi Caboni and Paola Scano

Received: 4 March 2021
Accepted: 13 May 2021
Published: 26 May 2021

Publisher's Note: MDPI stays neutral with regard to jurisdictional claims in published maps and institutional affiliations.

1. Introduction

Protected Designation of Origin (PDO) describes a group of agricultural/food products originating from a specific geographical area, for which production, processing and preparation must take place according to a recognised know-how [1].

In the European Union, 185 cheeses are registered as PDO products; this number rises to 231 if cheeses registered as Protected Geographical Indications (PGI) are considered [2]. A first classification of these products can be made based on milk processing. A total of 91 cheeses out of 231 (PDO + PGI) compulsorily require the use of raw milk; in other cases, the use of both raw and pasteurized/heat-treated milk is allowed. The lack of a specific requirement allows producers to freely decide how to process milk, leading to cheeses characterized by very different attributes [3].

The effects of milk heat treatment on cheese features have been extensively investigated and explained. From a microbiological and sensory point of view, the heat treatment depletes pathogens along with the characteristic microbiota, resulting in a more uniform final product that exhibits a less characteristic flavour profile compared to the raw milk counterpart [4]. From a technological point of view, though the application of a heat treatment guarantees improved hygienic conditions during processing and a better safety of the final product, it can result in an impaired coagulation and syneresis and in an alteration of cheese rheology and texture [5,6].

Fiore Sardo PDO (FS) cheese is the oldest cheese of Sardinia (Italy), being historically produced by shepherds in very small artisanal cheese factories. It must be obtained exclusively using raw whole milk from the Sarda Sheep breed. The official cheesemaking protocol still follows the ancient traditional process [7]. The peculiarities of the manufacturing process confer to Fiore Sardo a firm, crumbly and grainy texture, which easily breaks into flakes, and a savoury, piquant and smoky flavour [8–11]. Market offer is characterized by a quite heterogeneous variety of Fiore Sardo PDO, with very different quality attributes, with almost 90% of the product supplied by the dairy industry (Agris Sardegna, 2015).

In particular, the use of heat-treated milk for FS production has been associated with common industrial processing practices [12–15]. Previous studies concluded that this practice cannot be excluded [16] and that it is not unrealistic that some FS are manufactured using heat-treated milk, against the specifications indicated in the EC Regulation. From a scientific point of view, several chemical and physical characteristics of some productions were found to be reasonably explained by assuming the effects of thermal treatments on milk [16]. This is also supported by the fact that no official analytical methods able to discriminate between ewe milk cheese obtained from raw and thermised milk are available [17].

Some analytical techniques able to discriminate products made from raw milk and heat-treated milk have been proposed for product quality safeguards [17–21].

Analytical approaches based on nuclear magnetic resonance (NMR) relaxometry are particularly interesting. This technique was previously applied to describe the thermal denaturation of whey proteins and other milk components. Lambelet et al. [22], for example, demonstrated that irreversible thermal denaturation of whey proteins can be detected by low-field NMR. They observed that when NMR transverse relaxation rates were plotted versus heating temperature, a sigmoid curve was obtained, with relaxation rate values showing a steep increase from about 60 °C to about 80 °C, and a flex corresponding to milk thermisation/pasteurization temperatures usually adopted in dairy industrial processing (64 °C to 72 °C range). Later work by the same research group extended the results to other milk components and dyalized skim milk [23], suggesting the suitability of NMR relaxometry for measuring the thermal denaturation of milk proteins. Recent works, based on a high sample numerosity and exploiting an experimental setup more closely resembling real industrial processing conditions, demonstrated that the effect of heat treatment on milk is transferred to NMR relaxometry characteristics of both rennet curd and aged cheese. Raw and heat-treated milk [24] and derived curds [25,26] could be successfully discriminated by benchtop low-field NMR relaxometry. A similar discriminative ability of NMR relaxometry was also observed in ewe milk mature cheese by high-field magnetic resonance imaging (MRI) [19].

The aforementioned investigations demonstrated the sensitivity of NMR relaxation parameters to key cheesemaking processes (mainly thermal treatments of milk) in real dairy systems (whole milk, curds, mature cheese), suggesting the suitability of NMR relaxometry as a valid and useful tool in quality control activities in the dairy industry and for the safeguard of typical dairy productions.

In this study, heat treatments of milk were intentionally applied to produce FS cheese, and NMR relaxometry of raw and heat-treated FS were compared. Moreover, a comparison between artisanal and some industrial FS, the latter purchased from the market and visibly different from the typical FS PDO, was carried out for a more comprehensive evaluation of the effect of production chain on relaxometry features and aiming to optimize the exploitation of NMR relaxometry to assess the adherence to the official production protocol. MRI was used to compare heat-treated samples to their raw counterparts. A supplemental investigation was also carried out by exploiting image analysis protocols on a selected subset of MRI images and digital pictures in order to correlate process-induced changes with MRI image features and the visual appearance of samples.

2. Materials and Methods

Two sets of experiments were carried out, as summarized in Table S1. The first dataset (Dataset 1) was acquired by analysing samples made by different artisanal producers in different seasons. Additionally, a selection of industrial cheeses was purchased from the local market and subjected to the same analytical protocol used to characterize samples from Dataset 1.

2.1. Production of Fiore Sardo (FS) Cheese

2.1.1. Dataset 1

Five shepherds (S1, S2, S3, S4, S5) who were artisanal manufacturers of Fiore Sardo were selected for cheese sample production. Each shepherd produced a total of 20 cheese wheels in two different seasons (March–April 2019, Season 1; January–February 2020, Season 2; 100 cheese wheels in total for the two seasons). Ten wheels out of 20 were produced from raw milk, while the remaining ten were produced from the same milk after thermisation. Milk thermisation was carried out in different industrial plants, using plate heat exchanger modules. Cheesemaking was performed according to the official cheesemaking protocol [7,9], but milk thermisation was used for selected samples. Aged cheese samples were collected after 105 and 180 days of ripening in each shepherd's cellar. One quarter portion of each cheese wheel was sampled at each of the abovementioned time points, for a total of 10 wheels per producer per ripening time, 5 of which were made from raw milk (RC) and 5 from heat-treated milk (HTC).

2.1.2. Dataset 2

Industrial FS cheeses with similar ripening (6–8 months) were purchased from the local market in different seasons (January–November). Both cheeses manufactured by industrial dairy factories (11 samples) and from a maturer industry (2 samples) were purchased. Maturer industries buy Fiore Sardo from artisan producers and ripen these cheeses in their industrial cellars for several months, until they can be released to the market, i.e., after at least 105 days of ripening, according to the official cheesemaking protocol. Fiore Sardo PDO cheeses from the maturer industry are sold with the industrial brands, but milk is processed by shepherds and small artisans according to traditional processes [9]. A schematic representation of sample production and experimental plans (relative to both Dataset 1 and Dataset 2) is reported in Figure S1 (Supplementary Material).

2.2. Magnetic Resonance Imaging (MRI) Analysis

Each analytical sample of cheese subjected to MRI analysis represented the most central part of the cheese wheel, as schematically depicted in Figure S2.

In brief, each quarter was carefully cut in the portion corresponding to the wheel centre to obtain one cylindrical sample (diameter 2.3 cm, height 3–4 cm) to easily fit into a 50 mL Falcon tube. All samples were equilibrated to 22 °C for 1 h before MRI analysis. MRI experiments were performed using a Bruker Avance 300 MHz equipped with a microimaging Micro2.5 probe (Bruker Biospin, Karlsruhe, Germany) at room temperature (approximately 22–24 °C). A conventional Carr Purcell Meiboom Gill (CPMG) spin echo sequence was used with the following parameters: single 1 mm slice; echo time (TE) = 7.907 ms; repetition time (TR) = 3 s; echoes = 64; matrix = 128 × 128; number of excitations = 1–3; acquisition time = 25 min. To selectively analyse signals from water, the fat signal was suppressed, taking advantage of the different chemical shifts of fat and water at 7.05 T, with a 90° selective pulse applied before the conventional CPMG spin echo sequence at a frequency offset of 3.5 ppm with respect to water (preparatory fat saturation scheme provided by Bruker Topspin). All MR images were characterized by a signal to noise ratio higher than 100.

The AnalyzeNNLS software package was used to deconvolute the multi-exponential transversal relaxation time (T_2) decay of water proton signals [27]. Bruker MRI image data were converted into multiecho image data (MEID) files (Matlab, R2020b, The Mathworks, Sherborn, MA). Regions of interest (ROIs) were manually drawn on the MEID file (five ROIs for each image) and automatically processed to obtain the geometric mean T_2 (T_2 relaxation time constant, corresponding to the maximum intensity of the peak, on a logarithmic scale) and the area fraction (obtained by dividing the sum of the T_2 distribution within a desired region between T_2 min and T_2 max by the sum of all T_2 distributions).

2.3. Moisture Content

Moisture content was measured using the rapid determination method for cheese (130 °C, 90 min; [28]).

2.4. Statistical Analysis

The effect of thermisation was evaluated by comparing MRI parameters (geometric mean T_2 and area fractions) between RC and HTC by using univariate Student's *t*-test. An additional statistical comparison (one-way ANOVA, followed by Tukey post-hoc test) was performed to compare each shepherd's (i.e., artisanal) production to the industrial and maturer industry cheeses. Industrial cheeses (11 samples) were considered as a unique group of data, as were maturer cheeses (2 samples). Both tests were carried out with the MetaboAnalyst tool (https://www.metaboanalyst.ca, v.5, accessed 20 November 2020) [29], considering a significance threshold limit of $p < 0.05$. Box plots were obtained using MetaboAnalyst.

2.5. MRI Data Conversion

T_2 MRI Bruker Paravision raw 2dseq data from full Dataset 1 were first converted to Neuroimaging Informatics Technology Initiative (NIfTI, .nii) format using Bru2nii software (Bruker2NIfTI v1.0.20180303: by Matthew Brett, Andrew Janke, Mikaël Naveau, Chris Rorden, Windows 64-bit) with a resizing factor of 10. Then, the 64 slices from each NIfTI file were converted in JPEG format using a Matlab script first developed by Alex Laurence [30]. Only MRI images derived from the first MSME slice (TE = 7.907 ms) were used for deep transfer learning classification purposes.

2.6. Photographic Acquisition and Processing of Cheese Paste Surfaces

Snapshots of the surface of Fiore Sardo cheese from Dataset 1 produced in Season 1 at 105 days of ripening were taken using a Machine Vision System (MVS) assembled in our laboratory. Briefly, our MVS consisted of a portable plastified (PVC) white light-box (24 × 23 × 22 cm) equipped with a USB LED strip of white-light-emitting diode lamp (input 5 V, lumen 550, colour temperature 6500 K), assembled on a metallic support and directed upwards at a 90° angle with a cardboard box wall (Figure S3). Photos were taken without flash using a 25-megapixel-wide camera integrated in a Galaxy A50 smartphone. Cheese quarters were placed under a dark background and manually adjusted to be below the camera with a fixed distance of 25 cm. All operations were carried out in a dark room to minimize the effect of outside light. The acquired images (Joint Photographic Expert Group—JPEG; 24-bit colour depth, spatial resolution 3072 × 2048 pixels) were stored on a removable memory card. Images were finally processed by removing black background and manually cropped by using in-house-modified scripts available in the Matlab Image Processing Toolbox (Matlab, R2020b, The Mathworks, Sherborn, MA, USA).

Both digital pictures and MRI images were reordered in a concatenated folder structure according to the scheme illustrated in Figure S4. Each subfolder, related to the two ripening stages for each season, was separately compressed in zip format as input data for the deep learning classification described in the following section.

2.7. Deep Transfer Learning Based Classification

The proposed classification method relies on the deep transfer learning approach based on deep neural network (DNN) architectures. DNNs consist of a series of functions that take an input and return a predicted label. For this purpose, labeled input data were initially split into a training dataset (i.e., the set of data used to fit the model) and a validation dataset (i.e., the set of data used to provide an unbiased evaluation of a model fit on the training dataset while optimizing, or tuning, the model parameters). In brief, labeled input data pass through the DNN network using a loss function to evaluate how well the model performs at correctly identifying the true classes. The model optimizes for this loss function by computing the gradient of the loss function with respect to the model parameters. In parallel, the validation process optimizes the model parameters iteratively to minimize the loss and perform validation of the trained model [31]. Six pre-trained DNN frameworks were tested: ResNet [32], AlexNet [33], VggNet [34], SqueezeNet [35], DenseNet [36] and Inception-Net [37]. All models were implemented using the learning framework Pytorch, written in Python v. 3.7 and integrated into the open source GPU web server Google Colaboratory [38]. For the training step, all six DNN frameworks were trained under the following conditions: number of epoch = 1000; batch size = 8; learning rate = 0.001; momentum = 0.9. When the loss function converged and stabilized, the training step was automatically stopped and the training model saved. For the validation set, 20% of the training dataset data was used, and a 5-fold cross validation was performed to evaluate the performance of each model. Each model was trained 5 times on each dataset, and related performance parameters (training and validation accuracy and loss as well as training time) were computed. Finally, for all pre-trained models, the average values of the five replicates were calculated and listed in tables in order to compare classification performance.

3. Results and Discussion

3.1. MRI Analysis of Dataset 1

NMR relaxation of protons can occur only if stimulated by a fluctuating field of proper frequency that would induce the spin transition, in a non-spontaneous relaxation process that finally leads to equilibrium magnetization. Longitudinal relaxation, defined by the constant T_1, is triggered by field fluctuations at the Larmor frequency. Contrarily, in solid or semi-solid systems such as cheese, the major contribution of transverse relaxation, described by T_2, comes from the slow molecular dynamics of the studied liquid at zero Larmor frequency. For this reason, observed T_1 values usually present a very broad distribution, while populations are sharper and generally better defined for T_2. We chose the latter NMR relaxometry constant to describe the analysed cheeses.

Informative relaxometry features of the samples can be derived by analysing the multiexponential decay of the transverse magnetization from CPMG experiments at variable interpulse spacing [19,27,39]. Fitting of CPMG decays of MRI images of FS cheeses indicated a multiexponential behaviour (i.e., two T_2 populations), deriving from the presence of two water proton pools (Figure 1).

Multiexponential decay of the transverse component of NMR magnetization arises from the combined effect of diffusive and chemical exchange phenomena. In this condition, water molecules in cheese could be described, at a first approximation, as either free water or hydration water molecules (i.e., water embedded in the protein matrix and strongly interacting with it). Of course, the term "free" could be misleading, since water dynamics in this fraction are in fact much slower than in bulk water. NMR relaxometry is able to clearly detect the presence of different water pools, since water molecules sample different domains within the time scale of the CPMG experiment (i.e., the interpulse spacing). When diffusion is slow (and/or when relaxation domains are large), water molecules in the free water domain and hydration water retain their characteristic relaxation times, and multiexponential distribution is observed (e.g., in aged cheese); when diffusion is fast (and/or heterogeneity of domains is small) (e.g., in protein solutions, milk and curd), relaxation of water in different compartments is averaged, and proton transverse relaxation is mainly monoexponential [24,39,40].

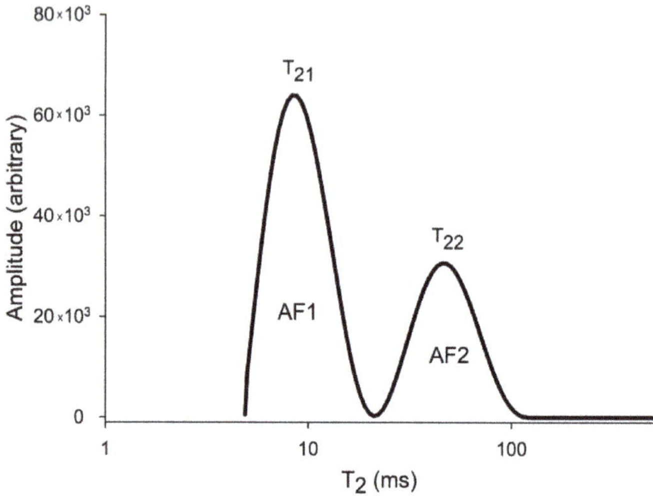

Figure 1. Representative T_2 distribution profile of an aged FS cheese at 300 MHz. Two proton populations are observed, characterized by geometric mean T_2 values (T_{21} and T_{22}) and area fractions of each population (AF1 and AF2).

At high field, diffusive exchange of water and fast chemical exchange between water protons, labile protein protons, and exchangeable protons in small molecules (such as sugars, vitamins, etc.) play a significant role in characterizing the observed relaxation distributions (Figure 1).

It was shown that in skim milk chemical exchange is mainly modulated by whey proteins [40], but in whole milk, the role of fat globules should be taken into account, since fat influences NMR relaxometry results [24]. In addition, it is clear that in cheese fat, protons might reasonably have a role in defining the observed NMR relaxometry profiles [41]. However, in the following discussion, the effect of fat protons on the observed T_2 distributions can be reasonably neglected, since fat signal has been saturated by a solvent suppression pulse in the MRI sequence, as previously explained [19].

For each pool, the mean area fractions (AF1 and AF2) and the corresponding T_2 values (i.e., the maximum intensity of each peak, T_{21} and T_{22}) were obtained (Table 1).

These data are in agreement with previous studies on aged ewe milk cheese and FS PDO [19], which also reported a first T_2 population relaxing in the range of 7–12 ms and a second one at 45–53 ms. The main novelty with respect to the mentioned data consists of the fact that the cheeses analysed in the present work were all made by following the official cheesemaking protocol of FS PDO, except for the thermal treatment of milk. Furthermore, a larger amount of data were used here to preliminarily validate observed data trends; the present results were indeed acquired during a much larger lag time (two production seasons), and two ripening times (105 and 180 days) were compared. Moreover, products from five different artisanal producers plus industrial Fiore Sardo PDO cheeses were analysed.

Table 1. T_{21} and T_{22} values (ms) and corresponding mean area fractions AF1 and AF2 (%) of artisanal FS cheeses produced in Season 1 and Season 2; asterisks (*) in each row indicate significant differences (*t*-test, $p < 0.05$) between RC and HTC parameters (T_{21} and T_{22}, AF1 and AF2) for each sample (M = Mean; SD = Standard Deviation).

| | | | T_{21} | | | | | T_{22} | | | | | AF1 | | | | | AF2 | | | | |
| | | | RC | | HTC | | | RC | | HTC | | | RC | | HTC | | | RC | | HTC | | |
			M	SD	M	SD		M	SD	M	SD		M	SD	M	SD		M	SD	M	SD	
Season 1	105 days	S1	11.16	0.47	9.48	0.40	*	52.58	3.70	56.01	1.56	*	0.75	0.07	0.85	0.02	*	0.24	0.07	0.14	0.02	*
		S2	11.28	0.94	9.24	0.42	*	49.05	3.84	52.15	1.85	*	0.70	0.03	0.78	0.03	*	0.32	0.05	0.21	0.03	*
		S3	10.74	0.78	9.13	0.58	*	52.37	3.54	47.51	0.77	*	0.82	0.02	0.82	0.02		0.17	0.02	0.18	0.02	
		S4	8.46	0.31	8.64	0.51		48.33	4.15	52.67	3.12	*	0.71	0.05	0.77	0.03	*	0.28	0.05	0.21	0.03	*
		S5	8.86	0.31	9.33	0.52	*	51.73	1.48	49.69	1.46	*	0.77	0.02	0.81	0.02	*	0.23	0.02	0.19	0.02	*
	180 days	S1	11.94	1.17	8.76	0.24	*	47.83	3.91	48.31	2.23		0.52	0.09	0.62	0.03	*	0.44	0.06	0.37	0.03	*
		S2	8.83	0.14	8.18	0.23	*	44.92	1.38	53.34	1.05	*	0.74	0.02	0.77	0.02	*	0.25	0.02	0.22	0.02	*
		S3	10.52	0.91	7.79	0.47	*	51.98	1.92	44.31	1.81	*	0.79	0.01	0.79	0.03		0.20	0.01	0.20	0.02	
		S4	9.74	1.64	7.98	0.28	*	40.67	3.29	47.27	2.17	*	0.56	0.03	0.70	0.03	*	0.42	0.08	0.29	0.03	*
		S5	8.50	0.31	8.80	0.74	*	49.42	1.49	46.66	1.93	*	0.67	0.03	0.69	0.05	*	0.32	0.03	0.30	0.05	*
Season 2	105 days	S1	12.70	1.60	12.70	0.69		52.47	2.85	57.72	2.80	*	0.68	0.05	0.78	0.04	*	0.31	0.05	0.22	0.04	*
		S2	14.11	1.45	12.23	0.93	*	54.40	1.59	53.04	2.98	*	0.63	0.06	0.78	0.04	*	0.36	0.06	0.21	0.04	*
		S3	10.26	0.26	10.02	0.39	*	45.93	1.05	52.22	1.46	*	0.79	0.01	0.84	0.03	*	0.20	0.02	0.16	0.03	*
		S4	10.20	1.48	10.49	0.31		46.95	3.28	50.16	2.18	*	0.67	0.07	0.79	0.01	*	0.32	0.07	0.20	0.01	*
		S5	9.29	1.08	10.74	0.42	*	43.55	3.08	53.18	1.05	*	0.68	0.02	0.82	0.01	*	0.31	0.02	0.17	0.01	*
	180 days	S1	9.08	0.18	8.82	0.50	*	49.26	2.67	52.37	4.38	*	0.62	0.04	0.78	0.06	*	0.37	0.04	0.21	0.06	*
		S2	9.94	0.66	9.21	0.44	*	53.10	2.35	52.33	1.93		0.72	0.05	0.80	0.03	*	0.27	0.05	0.19	0.03	*
		S3	10.39	0.50	8.78	0.35	*	50.97	1.01	52.90	1.61	*	0.72	0.03	0.76	0.03	*	0.27	0.03	0.23	0.03	*
		S4	9.09	0.74	8.31	0.51	*	50.87	3.87	48.03	3.28	*	0.68	0.05	0.73	0.03	*	0.31	0.05	0.26	0.03	*
		S5	14.55	2.24	10.07	0.38	*	44.94	3.34	53.00	1.14	*	0.57	0.03	0.72	0.02	*	0.42	0.03	0.27	0.02	*

Statistical analysis of MRI relaxometry values (Table 1) indicated significant differences in T_{21} and T_{22} values between RC and HTC cheeses. Evaluation of AF suggested the same consistent trend, with a significantly larger amount of protons in the faster relaxing population (AF1) in HTC cheeses than in RC. This is in quite good agreement with previous results on MRI relaxometry of heat-treated milk cheeses [19] and on the sensorial evaluation of Fiore Sardo cheese, suggesting that sensorial attributes and relaxometric features may correlate to each other and that they are both more appreciably affected by milk properties and cheesemaking processes than ripening [11].

Since for each production reported in Table 1 (S1–S5) the same milk was used, the manufacturing process was carried out by the same dairyman, and ripening time was exactly the same for both treatments (RC and HTC), the only factor that accounts for the observed relaxometry differences is milk thermisation.

From a previous investigation [19] and in accordance with pertinent literature [42–44], the fast relaxing proton pool (described by the pair T_{21}; AF1) can be associated with a fraction of water molecules strongly interacting with protons and entrapped in the para-casein network (protein hydration water); the slower population (T_{22}; AF2) can be associated with more mobile water molecules, which less strongly interact with the cheese protein network (which is sometimes referred to as bulk or "free" water).

Several different mechanisms may reasonably affect observed transverse relaxation rates. The first mechanism is related to an altered proportion between more mobile (free) water pools and water in the vicinity of the protein surface (hydration water), which is reasonably well described by AF1 [16,19,45]. This mechanism is in turn modulated by the diffusive exchange rate between water entrapped within the protein network of cheese, confined in cheese holes or pores and hydration water, and chemical exchange [39,46] between water protons and labile protein protons (mainly whey proteins [40]). The second mechanism involves internal water molecules in casein micelles, sometimes referred to as "bound" water molecules, which represent structural water molecules in the internal cavities, engaged in slowly modulated intermolecular dipole couplings with protein protons [46]. A significant effect of these "bound" water molecules on the observed proton transverse relaxation rate has been already suggested in dairy systems [40]. A third mechanism is explained by considering changes in the correlation time of macromolecules as a function of the ripening and processing methods [16,47]. Finally, relevant contribution could be expected from the mineral equilibrium of milk and its effect on micelles hydration. Equilibrium between colloidal (solid) and soluble calcium phosphate is in fact affected by temperature [48,49]; this has also been observed in Fiore Sardo cheeses manufactured from raw or heat-treated milk [6]. In particular, higher total calcium was found in Fiore Sardo produced from raw milk than from thermised milk, and it was hypothesized that mineral equilibrium in milk was shifted from soluble calcium to insoluble colloidal calcium as an effect of a high treatment temperature, leading to a higher retention of calcium in Fiore Sardo from raw milk [6]. Micellar calcium phosphate is a highly hydrated colloid, which greatly influences proton transverse relaxation. Ca and P association to casein and micelle hydration are strictly related phenomena, having a marked effect of NMR transverse relaxation of protons [40,48,49]. Heat treatment of milk was in fact found to cause precipitation of calcium phosphate and correspondingly affect proton T_2 [48].

3.2. MRI Analysis of Dataset 2

Industrial Fiore Sardo samples were characterized by the presence of two water proton pools, in agreement with results on artisanal samples RC and HTC discussed above. Mean area fractions (AF1 and AF2) and T_2 values (T_{21} and T_{22}) are shown in Table 2.

Both relaxation time constants T_2 and area values AF were consistent with the HTC artisanal cheeses analysed in Dataset 1, with the exception of samples from the maturer industry (I12, I13).

Both the increase of AF1 and the decrease of T_{21} are the result of a strong interaction between water protons and the para-casein protein network of cheese, i.e., of cheese protein hydration. From a certain point of view, AF1 more directly describes the level of protein hydration, being determined by the amount of protons in this pool. However, a decrease of the T_{21} value, being the result of the weighted average between the (long) T_2 of protons in free water pools and that (very short) of labile protein protons, characterized by a different chemical shift and affected by the typically long reorientational times of biological macromolecules, can be indirectly influenced by the amount of protein hydrating water.

Aiming to find suitable and objective descriptors of the relaxometric changes induced in cheese by milk thermisation, we define a new parameter, namely the Score ($) factor, as a representative indicator of the state of cheese microstructure. The Score ($) is defined as follows:

$$S = T_{21}/AF1 \tag{1}$$

where T_{21} and AF1 are the relaxation time and the area fraction of population 1, respectively, of the MRI T_2 distribution profile (Figure 1). Based on what was discussed above, the lower the Score, the more affected cheese microstructure is by milk thermisation. Correspondingly, the higher the Score, the less influenced are cheese proteins by any thermal effect on milk, i.e., milk can be considered as raw or untreated.

The calculated $ values (Season 1 and Season 2) are presented in Figure 2.

Even at first glance, two considerations can be made when observing Figure 2: first, thermisation of milk in HTC cheeses (red symbols) leads to a reduction of $ with respect to their raw RC counterparts (blue symbols); second, RC samples show much higher $ variability than their HTC counterparts.

Table 2. T_{21} and T_{22} values and correspondent mean area fractions AF1 and AF2 of industrial FS cheeses (M = Mean; SD = Standard Deviation).

	T_{21}		T_{22}		Area 1		Area 2	
	M	**SD**	**M**	**SD**	**M**	**SD**	**M**	**SD**
I1	8.301	0.055	40.337	0.100	0.835	0.002	0.157	0.002
I2	10.032	0.056	52.046	0.141	0.891	0.001	0.102	0.001
I3	9.214	0.038	45.078	0.035	0.814	0.001	0.178	0.001
I4	8.523	0.030	29.646	0.043	0.871	0.002	0.120	0.002
I5	8.018	0.024	51.936	0.047	0.697	0.002	0.294	0.002
I6	8.742	0.020	44.319	0.042	0.836	0.001	0.156	0.001
I7	9.318	0.023	51.352	0.133	0.858	0.012	0.130	0.000
I8	9.462	0.022	51.054	0.058	0.813	0.001	0.180	0.000
I9	9.463	0.018	51.058	0.037	0.813	0.001	0.180	0.001
I10	9.466	0.114	48.470	0.292	0.876	0.002	0.117	0.002
I11	9.188	0.047	46.314	0.131	0.843	0.001	0.150	0.001
I12	7.702	0.098	34.887	0.084	0.575	0.004	0.415	0.004
I13	7.301	0.029	34.772	0.013	0.545	0.002	0.442	0.002

Cheeses from one producer (S5) showed subtle, though significant, NMR relaxometry changes upon milk thermisation (blue and red circles are superimposed in Figure 2). In fact, for producer S5, both AF1 and T_{21} values always showed only slightly significant changes when RC and HTC cheese samples were compared at 105 days of ripening and at 180 days in Season 1 (Table 1). It should be considered that, at the shortest ripening time allowed for Fiore Sardo commercialization (105 days of ripening), the effect of water redistribution and protein hydration might still be not sufficient in order to clearly discriminate RC from HTC. In these cases, NMR relaxometry analysis should be replicated after a longer ripening time to obtain clearer results (Figure 2b,d). If a discrimination between RC and HTC is still not present at six months of ageing, then very likely milk has been overheated to some extent, and cheese features are not comparable to raw milk cheeses anymore. This result can be explained by assuming that, though thermisation was not intentionally performed in S5 production, raw milk was likely subjected to excessive heat before rennet addition. In fact, it is worth recalling here that, even at an artisanal level, milk is always treated to some extent during cheesemaking, since rennet is usually added when milk reaches, upon heating, a temperature in the range of approximately 35 and 38 °C, depending on the season, site, climate and usual practices of production. Among Fiore Sardo artisan producers, milk heating, according to traditional cheesemaking processes, is mostly carried out by directly warming up (often by means of an industrial stove gas cooker) a copper boiler containing raw milk. In some other cases, but still at artisanal level, more modern stainless steel multipurpose cheesemaking tanks are used to heat milk by means of water vapour in somewhat larger dairy plants. Only bigger industrial dairy plants adopt flow pasteurizers equipped with sophisticated plate heat exchanger modules, in order to perform more controlled high temperature–short time heat treatments. This consideration could explain the very subtle difference between the NMR relaxometry parameters of RC and HTC cheeses at 105 days and 180 days of ripening in S5 samples.

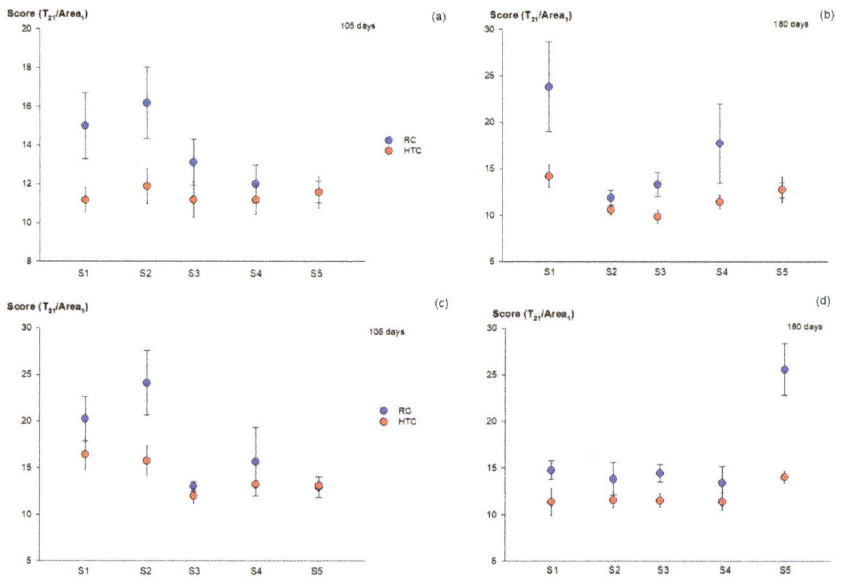

Figure 2. Score factors ($) of Fiore Sardo cheeses from Dataset 1: 105 days (**a**) and 180 days (**b**) of ripening, Season 1 (March–April 2019); 105 days (**c**) and 180 days (**d**) of ripening, Season 2 (January–February 2020). Blue circles represent samples made from raw milk (RC) and red circles represent cheeses made starting from heat-treated milk (HTC).

Notably, S5 and other shepherd's productions show superimposed error bars in Figure 2. This is mainly due to the large variability of RC and to similar technological considerations as already discussed for S5.

As explained above, the score value $ represents and describes the status of water molecules more tightly interacting with the paracasein network of cheese. Low score values (i.e., higher AF1 and/or lower T_{21}) indicate a high degree of protein hydration, arising from either ripening [16,45] or from the effect of heat treatments [19]. For cheeses having the same ripening time, the effect of the thermal treatment of milk on the following processing steps [5] dominates relaxation.

In fact, previous studies carried out at variable fields by fast field cycling NMR relaxometry (FFC-NMR) have demonstrated that the hydration of cheese proteins is affected by cheese ageing [45,47]. Further FFC-NMR studies have demonstrated that, for a given ageing time, cheese protein hydration can also be significantly affected by the different processing methods preceding cheese maturation [16]. As discussed above, two similar Fiore Sardo cheeses, which only differ in their heat treatment of milk, also differ in their NMR relaxometry characteristics. The reasons for the observed relaxometry differences can be found in cheese microstructure (such as fractal dimension, protein proton to water proton ratio, water mobility and protein hydration) [16].

The variability of Score values ($) within each group and among groups of cheeses is presented in the form of box plots in Figure 3. Standard deviations of RC of shepherds (blue boxplot) were generally larger than the other groups. Median values of RC are higher than all other groups, followed by maturer cheeses, HTC and industrial samples, with the latter showing the lowest median values. RC samples show considerably higher interquartile range than HTC and much larger whiskers. This could be reasonably ascribed to both the effect of the artisanal manufacturing process on cheese microstructure and to the microbiological complexity of the systems arising from the use of raw milk. In fact, while artisanal cheesemaking entails mainly manual practices, industrial production implies more standardized protocols and a higher degree of process automation. In addition, previous reports on raw and heat-treated milk cheeses demonstrated a larger variability of raw-milk cheeses than pasteurized-milk cheese counterparts, and it was suggested that the effect of heat on milk proteins, indigenous enzymes and microbial flora of milk may reduce variability, resulting in a more homogeneous peptide profile, sensory properties and NMR relaxation parameters for pasteurized-milk cheese [19,50,51].

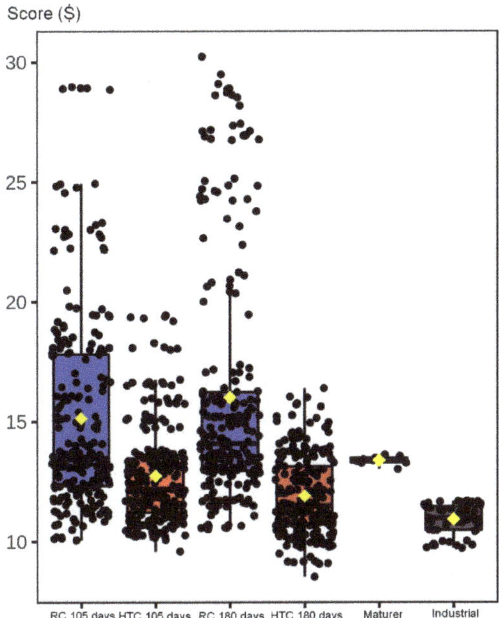

Figure 3. Box plots of the Score values ($) of each group of cheeses (RC, HTC, industrial and maturer cheeses).

Differences in the manufacturing process are reflected in different physico-chemical and sensory features, resulting in the distinctiveness of each wheel from the same shepherd. While heating milk to 35–38 °C, necessary for a proper action of rennet, is usually not strictly standardized in terms of temperature control, industrial processing is carefully controlled. Mild thermal treatments of raw milk, typical of artisanal practices, allow for the preservation of the microorganisms that represent a sort of fingerprint of each shepherd's production. The development of indigenous milk microflora and the related molecular changes (e.g., changes in pH, proteolysis and lipolysis during ripening, etc. [5]) occurring during the following processing steps, together with the exclusively manual manufacturing operations on curd, are reflected in the microstructural, molecular and sensory properties of the final cheese, making every wheel a unique product. That is why artisanal Fiore Sardo PDO cheeses are very different from each other, being strongly linked to the territory and to the specific production practices. It is interesting to note, in this regard, that milk and curd handling practices for making Fiore Sardo are passed down from generation to generation and often differ from each other in subtle, but likely relevant, details. On the other hand, when heat is applied to the same milk, though (almost) the same manufacturing practices are carried out by the same cheesemongers, part of the characteristic microflora is depleted and milk functionality is altered [51,52]. The $ data distribution in HTC cheeses indeed indicated a reduced heterogeneity in comparison to RC. The box plot of industrial cheeses more clearly shows that the spreading of $ values is considerably reduced and very narrow in comparison to both RC and HTC. This is mostly ascribable to industrial practices, where cheesemaking is strictly controlled and standardized in terms of both microbiological conditions and technological operations. It is worth noting that industrial Fiore Sardo samples were purchased from different industrial producers, so the observed low interquartile range should be considered quite representative of the standardization of the industrial production in the market and cannot be ascribed to a specific producer.

Mean and standard deviations of Score values $ for all groups of samples (RC, HTC, I and maturer), together with their statistical significances, are presented in Table 3.

Table 3. Statistical comparison of Score ($) values (ANOVA, $p < 0.05$) of the four groups of Fiore Sardo samples, namely RC, HTC (produced in Season 1 and Season 2), maturer and industrial cheeses. Each shepherd production (RC and HTC) was individually compared to maturer and industrial cheeses; data are expressed as Mean ± Standard Deviation; means sharing the same superscript letters are not significantly different from each other; the letter "a" is assigned to the higher mean.

			RC	HTC	Maturer	Industrial
S1	Season 1	105 days	14.99 ± 1.70 [a]	11.17 ± 0.62 [c]	13.39 ± 0.18 [b]	10.92 ± 0.62 [c]
		180 days	23.83 ± 4.83 [a]	14.22 ± 1.22 [b]	13.39 ± 0.18 [b]	10.92 ± 0.62 [c]
	Season 2	105 days	20.25 ± 2.38 [a]	16.43 ± 1.60 [b]	13.39 ± 0.18 [c]	10.92 ± 0.62 [d]
		180 days	14.78 ± 1.00 [a]	11.38 ± 1.45 [c]	13.39 ± 0.18 [b]	10.92 ± 0.62 [c]
S2	Season 1	105 days	16.18 ± 1.85 [a]	11.87 ± 0.90 [c]	13.39 ± 0.18 [b]	10.92 ± 0.62 [d]
		180 days	11.90 ± 0.81 [b]	10.57 ± 0.48 [c]	13.39 ± 0.18 [a]	10.92 ± 0.62 [c]
	Season 2	105 days	24.10 ± 3.44 [a]	15.74 ± 1.59 [b]	13.39 ± 0.18 [c]	10.92 ± 0.62 [d]
		180 days	13.86 ± 1.74 [a]	11.58 ± 0.86 [b]	13.39 ± 0.18 [a]	10.92 ± 0.62 [c]
S3	Season 1	105 days	13.12 ± 1.19 [a]	11.19 ± 0.90 [b]	13.39 ± 0.18 [a]	10.92 ± 0.62 [b]
		180 days	13.33 ± 1.29 [a]	9.86 ± 0.73 [c]	13.39 ± 0.18 [a]	10.92 ± 0.62 [b]
	Season 2	105 days	13.02 ± 0.51 [a]	11.98 ± 0.82 [b]	13.39 ± 0.18 [a]	10.92 ± 0.62 [c]
		180 days	14.47 ± 0.94 [a]	11.53 ± 0.70 [c]	13.39 ± 0.18 [b]	10.92 ± 0.62 [d]
S4	Season 1	105 days	12.00 ± 0.97 [b]	11.18 ± 0.75 [c]	13.39 ± 0.18 [a]	10.92 ± 0.62 [c]
		180 days	17.74 ± 4.27 [a]	11.48 ± 0.75 [bc]	13.39 ± 0.18 [b]	10.92 ± 0.62 [c]
	Season 2	105 days	15.65 ± 3.68 [a]	13.23 ± 0.40 [b]	13.39 ± 0.18 [b]	10.92 ± 0.62 [c]
		180 days	13.41 ± 1.77 [a]	11.42 ± 0.96 [b]	13.39 ± 0.18 [a]	10.92 ± 0.62 [b]
S5	Season 1	105 days	11.58 ± 0.56 [b]	11.57 ± 0.81 [b]	13.39 ± 0.18 [a]	10.92 ± 0.62 [c]
		180 days	12.74 ± 0.81 [a]	12.76 ± 1.39 [a]	13.39 ± 0.18 [a]	10.92 ± 0.62 [b]
	Season 2	105 days	12.93 ± 1.10 [a]	13.08 ± 0.53 [a]	13.39 ± 0.18 [a]	10.92 ± 0.62 [b]
		180 days	25.61 ± 2.79 [a]	14.05 ± 0.69 [b]	13.39 ± 0.18 [b]	10.92 ± 0.62 [c]

In general, ANOVA results suggested a consistent differentiation among the four groups of cheeses. Industrial Fiore Sardo showed the lowest $ values in all comparisons, except for one case (i.e., S3 180 days, Season 1), which had $ values higher than HTC only. RC often had significantly higher $ values than other groups of cheeses, except for two cases (i.e., S4 and S5, 105 days, Season 1), when maturer cheeses showed higher $. This result can be explained by considering that maturer cheeses were purchased after approximately 6–8 months of ripening. In some other cases, $ values were comparable for RC, HTC and maturer cheeses. Statistical comparison of HTC and industrial cheeses usually resulted in a significantly lower $ for the latter, but they were comparable in five cases. Maturer samples were never comparable to industrial samples.

Based on the aforementioned statistical evaluations, some conclusions can be drawn on the manufacturing process of each group of samples and on the quality features of the resulting cheeses. RC and HTC were entirely produced following artisanal procedures. Maturer cheeses share the same artisanal processing but were ripened in industrial-like conditions, in refrigerated cellars with controlled humidity and temperature. Despite different ripening conditions, RC and maturer cheeses were described by very similar $ values, clearly suggesting that technological transformations that precede ripening confer to cheeses a common relaxometric behaviour described by comparable $. RC and maturer cheeses were generally described by higher $ than industrial cheeses, although ANOVA suggested that RC, HTC and maturer cheeses can still differ to some extent among each other in terms of the new quality-related parameter, $. Since the RC and HTC cheeses of each shepherd were technologically processed in the same way, the decrease in $ observed between RC and HTC (Figure 3), supported by ANOVA comparisons (Table 3), can be reasonably and solely related to the application of a heat treatment of the milk. On the other hand, industrial samples, besides having a much narrower $ variability (Figure 3) clearly exhibited considerably and significantly lower $ values with respect to the other three groups of cheeses. These relaxometric behaviours could be reasonably associated with the features of industrial productions, for which milk and curd processing and handling are more mechanized and standardized, ensuring more uniform features in the final products, as previously reported [16]. Moreover, a contribution to the low $ may arise from the application of a heat treatment to the milk that is consistent with common industrial processing practices, as realistically suggested [12–15]. Interestingly, the discrimination ability of $ and its potential in identifying relevant quality features of Fiore Sardo cheeses does not seem to be affected by different seasons of lactation (i.e., to milk compositional quality), being instead more strictly related to milk processing. This consideration derives from the significant statistical differences among the four groups of samples, regardless the season (Figure 3, Table 3). As far as ripening conditions are concerned, the significant role of ripening cannot be excluded. However, the effect of ripening conditions seems to be less influential than the milk and curd processing practices preceding cheese ageing. This makes NMR relaxometry a suitable analytical tool for characterizing the effect of heat treatments of milk on quality features of Fiore Sardo cheese.

3.3. Moisture Content

The moisture content of all cheese samples are reported in Table S2.

Moisture content was not influenced by heat treatment, indeed showing no consistent trend in RC and HTC samples. This is in contrast to MRI parameters, which showed consistent changes according to heat treatments. Since MRI parameters T_{21} and AF1 could be considered representative of the portion of water in the samples, such changes could not be related to total moisture content [19]. In addition, for industrial samples, no noticeable correlation was found between MRI parameters and moisture content.

Water status can be described at different scales of investigation—at a macroscopic level by moisture content and at a molecular level by MRI—and can lead to different information about the analyzed matrix. Moisture content is representative of the extractable water molecules by drying in certain conditions and is among the most common parameters used to describe food quality. However, it should not be considered as an exhaustive parameter as it does not give any information on water proton pools in the system (if any) and their interactions with the protein matrix in cheese [53].

3.4. Image Analysis of Fiore Sardo Pictures and MRI Images

Representative pictures of the visual appearance of RC and HTC at the minimum allowed ripening time for Fiore Sardo PDO (i.e., 105 days) are shown in Figure 4.

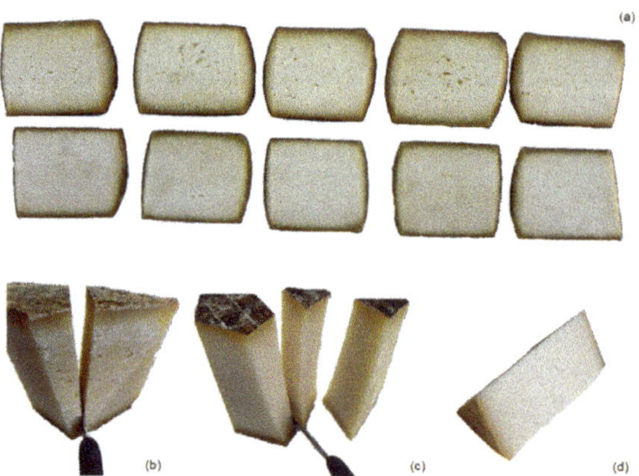

Figure 4. Representative images of the visual appearance of RC, HTC and industrial samples: Producer S1, Season 2, 105 days of ripening, Fiore Sardo from raw milk (upper row) and from heat-treated milk (lower row) (**a**); details of RC (**b**) and HTC (**c**) samples ripened 105 days upon cutting; industrial Fiore Sardo purchased from a local market, with at least 105 days of ripening (**d**).

Even at first sight, RC, HTC and industrial cheeses appear very different from each other. The HTC samples surface appears more homogeneous, smoother and with fewer holes, while RC cheeses are more grainy, friable and show more and bigger holes. When cut with a knife, RC cheeses are prone to easily break and crumble, while HTC appear more creamy and can be cut very easily and without breaking into pieces.

The different visual appearances of HTC samples with respect to RC cheeses is reasonably ascribable to the modifications resulting from the heat treatment of milk, which is known to have, in Fiore Sardo PDO, multiple consequences on cheese composition, microstructure and texture, which in turn affect appearance and sensory properties in general [6,10,52].

In general, MRI images confirmed, by a non-invasive approach, what already appeared at a first visual inspection of samples. RC cheeses exhibited a coarser paste, often with multiple cracks and holes, while HTC samples exhibited a smoother and more homogenous paste (Figure S5).

Two kinds of scientific considerations can be made based on the MRI images acquired. On one hand, as discussed above, MRI images give relaxometric information that is able to provide useful insights on the state of biopolymers and on the mobility of the system at molecular level. On the other hand, important information can be derived from the texture of MRI images by performing a detailed image analysis.

In recent years, image analysis techniques rapidly emerged not only at the academic level but also as rapid, economic, consistent and objective inspection techniques in the agricultural and food industry [54,55]. In particular, for the dairy industry sector, modern computer vision and image analysis methods have been used to evaluate cheese external or internal quality attributes such as colour, shape, texture, amount and distribution of added ingredients or production defects [56–58].

In this study, we pave the way for a deep-learning-based method for the classification of Fiore Sardo cheese images. The method consists of using data collected both from MRI and from a computer vision system based on photographic images taken with a smartphone camera. Deep learning is a subfield of artificial intelligence (AI) concerned with specific algorithms for thinking simulations and data processing inspired by the neurons of the brain [31]. Deep-learning-based image classification has proven to be highly effective in image classifying, object detection and other computer vision related problems [59,60].

All pre-trained DNN models run well in the open web GPU_Google colab service, with compatible and relatively short training times. For both MRI and photo datasets, the best classification result was achieved by the SqueezeNet model (Tables S3 and S4). In particular, for the T_2 weighted MRI images, the SqueezeNet model exhibits a training accuracy from 93% to 98%, with the highest validation score (100%) (Table S3). As far as the photo dataset is concerned, SqueezeNet shows a training accuracy of 96%, with the same validation of 100% (Table S4).

The other pre-trained models (ResNet, AlexNet, VggNet, Densenet and Inception-net) also exhibit satisfactory performance results, ranging from a minimum of 72% to a maximum of 92%, with related validation ranging from 77% to 100% (Tables S3 and S4).

We propose here an IA approach to classify photos and T_2 weighted MRI images of Fiore Sardo cheese made with raw or thermised milk, by applying a deep neural network (DNN), which includes several chained layers of processing between the input (photos or MRI images) and the output (classification/label of the input image).

As with most state-of-the-art IA methods, deep learning models usually require fitting thousands or millions of input data; thus, their application to those studies where the sample size is limited appears prohibitive. To address this challenge, the deep transfer learning approach offers a more suitable alternative. Transfer learning is the method by which the model uses the knowledge gained during the training of a relatively large dataset in a different but related problem (i.e., image classification and recognition). During the transfer learning process, only the selected classifier(s) is trained in the new network, while the features learned from the large dataset are transferred. Therefore, with the deep transfer learning approach, we do not need to retrain the entire network for a new dataset, thus allowing the training of deep learning models with relatively few data, reduced computational power and less training time [61].

With the main disadvantage of the present study being the relatively low number of available cheese samples, compared to the large data size usually needed for training deep learning frameworks, we chose to transfer the learning technique by testing six pre-trained DNN frameworks (ResNet, AlexNet, VggNet, SqueezeNet, DenseNet and Inception-Net), a widely used approach for addressing classification and image recognition challenges of MRI data.

In the end, all tested pre-trained DNNs yielded promising performances, giving an average classification accuracy of over 72% and over 93% in particular for the best performing SqueezeNet model (Tables S3 and S4). We did not perform further accuracy tests of the DNN pre-trained model in this study; however, we are proposing to implement and optimize this model in a dedicated study when more data are available.

4. Conclusions

Our data demonstrate that NMR relaxometry is able to assess molecular mobility changes induced in cheese by the most common industrial thermal treatments carried out on milk. In the present work, the experimental plan was optimized to exclude possible confounding factors. The same milk was used for producing both raw-milk or thermised-milk cheese. Moreover, analyses were repeated in different seasons, and the products from different producers were compared. Results were compared with a selection of industrial Fiore Sardo cheeses purchased in the local market.

NMR relaxometry data of Fiore Sardo cheese indicated the presence of two water populations, described by characteristic T_2 relaxation times and area fractions. In particular, the first and fastest relaxing population showed a good predictive value. In order to obtain a relaxometric descriptor of FS cheese microstructure and molecular mobility, a new quality-related parameter was introduced, namely the Score factor ($). The $ describes cheese in relation to the hydration status of the casein network and the state of the water-fat-protein network in cheese. We were able to discriminate FS cheeses produced from raw (RC) and heat-treated milk (HTC), with the latter exhibiting a lower $, indicating an enhanced hydration of the casein network. Industrially manufactured FS cheeses had generally lower $ values than RC and HTC. Samples from the maturer industry, which buys cheeses before ripening and ages cheeses in industrial-like refrigerated cellars, showed similar values to artisanal cheeses, suggesting that NMR relaxometry is more sensitive to changes occurring in the cheesemaking steps preceding ripening. A further discriminative approach was finally carried out by processing acquired MRI images and photos with an IA method based on pre-trained deep learning algorithms. Preliminary classification results suggested that this innovative approach is promising in providing suitable discrimination between RC and HTC Fiore Sardo cheese.

Supplementary Materials: The following are available online at https://www.mdpi.com/article/10.3390/dairy2020023/s1, Figure S1: sample production and sampling scheme for MRI analysis, Figure S2: MRI sample preparation scheme, Figure S3: Computer Vision System (CVS) for photographic analysis, Figure S4: Image Dataset Directory and File Structure used, Figure S5: Representative T_2 weighted MRI images of Fiore Sardo cheeses, Table S1: Production of FS cheeses and datasets, Table S2: Moisture content data of cheeses, Table S3: Performance and classification results of pre-trained DL frameworks on MRI images, Table S4: Performance and classification results of pre-trained DL frameworks on photographic images.

Author Contributions: Conceptualization, R.A.; Methodology, R.A., E.C. and R.M.; Software, E.C. and R.M; Formal Analysis, R.A., E.C. and R.M; Investigation, R.A., E.C. and R.M.; Data Curation, E.C. and R.M.; Writing—Original Draft Preparation, R.A., E.C. and R.M.; Writing—Review and Editing, R.A., E.C. and R.M.; Visualization, E.C. and R.M.; Supervision, R.A.; Funding Acquisition, R.A.; Project Scientific Coordination, R.A. All authors have read and agreed to the published version of the manuscript.

Funding: This research was funded by the Sardinia Regional Government by means of Sardegna Ricerche (art 9, L.R. n. 15, 17.11.2010).

Data Availability Statement: The data presented in this study are available on request from the corresponding author.

Conflicts of Interest: The authors declare no conflict of interest. The funders had no role in the design of the study; in the collection, analyses, or interpretation of data; in the writing of the manuscript, or in the decision to publish the results.

References

1. European Commission. Quality Schemes Explained. Available online: https://ec.europa.eu/info/food-farming-fisheries/food-safety-and-quality/certification/quality-labels/quality-schemes-explained (accessed on 16 February 2021).
2. European Commission eAmbrosia—The EU Geographical Indications Register. Available online: https://ec.europa.eu/info/food-farming-fisheries/food-safety-and-quality/certification/quality-labels/geographical-indications-register/# (accessed on 11 April 2021).
3. Sardo, P.; Barletta, M. European Designations between Identity Values and Market. 2019. Available online: https://www.slowfood.com/wp-content/uploads/2019/09/European-designations-between-identity-values-and-market_ENG.pdf (accessed on 26 February 2021).
4. Fox, P.; McSweeney, P.; Cogan, T.; Guinee, T. *Cheese—Chemistry, Physics and Microbiology*, 3rd ed.; Guinee, P., Fox, P., McSweeney, T., Cogan, T., Eds.; Academic Press: Cambridge, MA, USA, 2004; ISBN 978-1-4613-6138-1.
5. Fox, P.; Uniacke-Lowe, T.; McSweeney, P.; O'Mahony, J. (Eds.) Heat-Induced Changes in Milk. In *Dairy Chemistry and Biochemistry*; Springer International Publishing: Basel, Switzerland, 2015; pp. 345–375. ISBN 978-3-319-14891-5.
6. Pirisi, A.; Pinna, G.; Papoff, C.M. Effect of milk thermisation on Fiore Sardo PDO cheese: 1. Physicochemical characteristics. *Sci. Tecn. Latt. Cas.* **1999**, *50*, 353–366.
7. EUR-Lex—31996R1236—EN—EUR-Lex. Available online: https://eur-lex.europa.eu/legal-content/IT/TXT/?uri=CELEX%3A31996R1236 (accessed on 26 February 2021).

8. Anedda, R. Magnetic Resonance Analysis of Dairy Processing Suitable Tools for the Dairy Industry. In *Magnetic Resonance in Food Science: Defining Food by Magnetic Resonance*; The Royal Society of Chemistry: London, UK, 2015; pp. 49–64. ISBN 978-1-78262-031-0.

9. Anedda, R.; Pala, M.; Curti, E. Cheeses That Made History in Italian Dairy Tradition. In *Cheeses around the World: Types, Production, Properties and Cultural and Nutritional Relevance*; Ferrão, A.C., Guiné, R.P.F., Correia, P.R., Eds.; Nova Science Publishers: New York City, NY, USA, 2019; pp. 203–228. ISBN 978-1-53615-419-1.

10. Mendia, C.; Ibañez, F.C.; Torre, P.; Barcina, Y. Effect of pasteurization on the sensory characteristics of a ewe's-milk cheese. *J. Sens. Stud.* **1999**, *14*, 415–424. [CrossRef]

11. Scintu, M.F.; Del Caro, A.; Urgeghe, P.P.; Piga, C.; Di Salvo, R. Sensory profile development for an Italian PDO ewe's milk cheese at two different ripening times. *J. Sens. Stud.* **2010**, *25*, 577–590. [CrossRef]

12. Siniscalchi, V. Environment, regulation and the moral economy of food in the Slow Food movement. *J. Polit. Ecol.* **2013**, *20*, 295–305. [CrossRef]

13. Pittalis, P.; Cappellacci, U.; Contu, M.I.; Fasolino, G.; Peru, A.; Tedde, M.; Tocco, E.; Tunis, S.; Zedda, A. XV Legislature—Regional Question N.1380. Available online: http://www3.consregsardegna.it/XVlegislatura/interrogazioni/1380 (accessed on 26 February 2021).

14. Ledda A XV Legislature—Regional Question N.1706/A. Available online: http://www3.consregsardegna.it/XVlegislatura/interrogazioni/1706 (accessed on 26 February 2021).

15. Siniscalchi, V. Food, Slow Food and Middle Class Activism. In *Food Activism: Agency, Democracy and Economy*; Bloomsbury Academic: London, UK, 2014; pp. 9–24.

16. Anedda, R.; Pardu, A.; Korb, J.P.; Curti, E. Effect of the manufacturing process on Fiore Sardo PDO cheese microstructure by multi-frequency NMR relaxometry. *Food Res. Int.* **2021**, *140*, 110079. [CrossRef]

17. Caboni, P.; Maxia, D.; Scano, P.; Addis, M.; Dedola, A.; Pes, M.; Murgia, A.; Casula, M.; Profumo, A.; Pirisi, A. A gas chromatography-mass spectrometry untargeted metabolomics approach to discriminate Fiore Sardo cheese produced from raw or thermized ovine milk. *J. Dairy Sci.* **2019**, *102*, 5005–5018. [CrossRef]

18. Dedola, A.S.; Piras, L.; Addis, M.; Pirisi, A.; Piredda, G.; Mara, A.; Sanna, G. New analytical tools for unmasking frauds in raw milk-based dairy products: Assessment, validation and application to fiore sardo PDO cheese of a RP-HPLC method for the evaluation of the α-l-fucosidase activity. *Separations* **2020**, *7*, 40. [CrossRef]

19. Mulas, G.; Roggio, T.; Uzzau, S.; Anedda, R. A new magnetic resonance imaging approach for discriminating Sardinian sheep milk cheese made from heat-treated or raw milk. *J. Dairy Sci.* **2013**, *96*, 7393–7403. [CrossRef] [PubMed]

20. Mazza, M.; Guglielmetti, C.; Brusadore, S.; Sciuto, S.; Esposito, G.; Caramelli, M.; Peletto, S.; Acutis, P.L.; Marengo, E.; Manfredi, M.; et al. A proteomic approach to the safeguard of a typical agri-food product: Fiore sardo PDO. *Adv. Dairy Res.* **2019**, *7*. [CrossRef]

21. Piga, C.; Urgeghe, P.P.; Piredda, G.; Scintu, M.F.; Di Salvo, R.; Sanna, G. Thermal inactivation and variability of γ-glutamyltransferase and α-l-fucosidase enzymatic activity in sheep milk. *LWT Food Sci. Technol.* **2013**, *54*, 152–156. [CrossRef]

22. Lambelet, P.; Berrocal, R.; Ducret, F. Low resolution NMR spectroscopy: A tool to study protein denaturation: I. Application to diamagnetic whey proteins. *J. Dairy Res.* **1989**, *56*, 211–222. [CrossRef]

23. Lambelet, P.; Berrocal, R.; Renevey, F. Low-field nuclear magnetic resonance relaxation study of thermal effects on milk proteins. *J. Dairy Res.* **1992**, *59*, 517–526. [CrossRef]

24. Curti, E.; Pardu, A.; Melis, R.; Addis, M.; Pes, M.; Pirisi, A.; Anedda, R. Molecular mobility changes after high-temperature, short-time pasteurization: An extended time-domain nuclear magnetic resonance screening of ewe milk. *J. Dairy Sci.* **2020**, *103*, 9881–9892. [CrossRef]

25. Curti, E.; Pardu, A.; Del Vigo, S.; Sanna, R.; Anedda, R. Non-invasive monitoring of curd syneresis upon renneting of raw and heat-treated cow's and goat's milk. *Int. Dairy J.* **2019**, *90*. [CrossRef]

26. Curti, E.; Pardu, A.; Del Vigo, S.; Sanna, R.; Anedda, R. A low-field Nuclear Magnetic Resonance dataset of whole milk during coagulation and syneresis. *Data Br.* **2019**, *26*. [CrossRef] [PubMed]

27. Bjarnason, T.A.; Mitchell, J.R. AnalyzeNNLS: Magnetic resonance multiexponential decay image analysis. *J. Magn. Reson.* **2010**, *206*, 200–204. [CrossRef] [PubMed]

28. AOAC International. AOAC 948.12-2002, Moisture in Cheese. Method II (Rapid Screening Method): AOAC Official Method. Available online: http://www.aoacofficialmethod.org/index.php?main_page=product_info&products_id=1145 (accessed on 26 February 2021).

29. Pang, Z.; Chong, J.; Li, S.; Xia, J. MetaboAnalystR 3.0: Toward an optimized workflow for global metabolomics. *Metabolites* **2020**, *10*, 186. [CrossRef]

30. Laurence, A. NIfTI Image Converter (nii2png) for Python and Matlab | NIfTI-Image-Converter. Available online: https://alexlaurence.github.io/NIfTI-Image-Converter/ (accessed on 26 February 2021).

31. Shrestha, A.; Mahmood, A. Review of deep learning algorithms and architectures. *IEEE Access* **2019**, *7*, 53040–53065. [CrossRef]

32. He, K.; Zhang, X.; Ren, S.; Sun, J. Deep residual learning for image recognition. In Proceedings of the 2016 IEEE Conference on Computer Vision and Pattern Recognition (CVPR), Las Vegas, NV, USA, 27–30 June 2016; pp. 770–778.

33. Krizhevsky, A.; Sutskever, I.; Hinton, G.E. ImageNet classification with deep convolutional neural networks. In Proceedings of the 25th International Conference on Neural Information Processing Systems, Siem Reap, Cambodia, 13–16 December 2018; Curran Associates Inc.: Red Hook, NY, USA, 2012; Volume 1, pp. 1097–1105.

34. Liu, S.; Deng, W. Very deep convolutional neural network based image classification using small training sample size. In Proceedings of the 2015 3rd IAPR Asian Conference on Pattern Recognition (ACPR), Kuala Lumpur, Malaysia, 3–6 November 2015; pp. 730–734.

35. Iandola, F.N.; Moskewicz, M.W.; Ashraf, K.; Han, S.; Dally, W.J.; Keutzer, K. SqueezeNet: AlexNet-level accuracy with 50x fewer parameters and <1MB model size. *arXiv* **2016**, arXiv:1602.07360.

36. Huang, G.; Liu, Z.; Van Der Maaten, L.; Weinberger, K.Q. Densely connected convolutional networks. In Proceedings of the 2017 IEEE Conference on Computer Vision and Pattern Recognition (CVPR), Honolulu, HI, USA, 21–26 July 2017; pp. 2261–2269.

37. Szegedy, C.; Vanhoucke, V.; Ioffe, S.; Shlens, J.; Wojna, Z. Rethinking the inception architecture for computer vision. In Proceedings of the 2016 IEEE Conference on Computer Vision and Pattern Recognition (CVPR), Las Vegas, NV, USA, 27–30 June 2016; pp. 2818–2826.

38. Google Google Colaboratory. Available online: https://colab.research.google.com/notebooks/intro.ipynb (accessed on 26 February 2021).

39. Hills, B.P.; Takacs, S.F.; Belton, P.S. A new interpretation of proton NMR relaxation time measurements of water in food. *Food Chem.* **1990**, *37*, 95–111. [CrossRef]

40. Mariette, F.; Tellier, C.; Brule, G.; Marchal, P. Multinuclear Nmr-Study of the Ph dependent water state in skim milk and caseinate solutions. *J. Dairy Res.* **1993**, *60*, 175–188. [CrossRef]

41. Mariette, F. NMR Relaxation of Dairy Products. In *Modern Magnetic Resonance*; Webb, G.A., Ed.; Springer: Dordrecht, The Netherlands; Berlin, Germany, 2006; pp. 1697–1701. ISBN 978-1-4020-3894-5.

42. Gianferri, R.; D'Aiuto, V.; Curini, R.; Delfini, M.; Brosio, E. Proton NMR transverse relaxation measurements to study water dynamic states and age-related changes in Mozzarella di Bufala Campana cheese. *Food Chem.* **2007**, *105*, 720–726. [CrossRef]

43. Gianferri, R.; Maioli, M.; Delfini, M.; Brosio, E. A low-resolution and high-resolution nuclear magnetic resonance integrated approach to investigate the physical structure and metabolic profile of Mozzarella di Bufala Campana cheese. *Int. Dairy J.* **2007**, *17*, 167–176. [CrossRef]

44. Chaland, B.; Mariette, F.; Marchal, P.; De Certaines, J. ^{1}H nuclear magnetic resonance relaxometric characterization of fat and water states in soft and hard cheese. *J. Dairy Res.* **2000**, *64*, 609–618. [CrossRef]

45. Godefroy, S.; Korb, J.-P.; Creamer, L.K.; Watkinson, P.J.; Callaghan, P.T. Probing protein hydration and aging of food materials by the magnetic field dependence of proton spin-lattice relaxation times. *J. Colloid Interface Sci.* **2003**, *267*, 337–342. [CrossRef]

46. Venu, K.; Denisov, V.P.; Halle, B. Water ^{1}H magnetic relaxation dispersion in protein solutions. A quantitative assessment of internal hydration, proton exchange, and cross-relaxation. *J. Am. Chem. Soc.* **1997**, *119*, 3122–3134. [CrossRef]

47. Conte, P.; Cinquanta, L.; Lo Meo, P.; Mazza, F.; Micalizzi, A.; Corona, O. Fast field cycling NMR relaxometry as a tool to monitor Parmigiano Reggiano cheese ripening. *Food Res. Int.* **2021**, *139*, 109845. [CrossRef]

48. Boiani, M.; Fenelon, M.; FitzGerald, R.J.; Kelly, P.M. Use of 31P NMR and FTIR to investigate key milk mineral equilibria and their interactions with micellar casein during heat treatment. *Int. Dairy J.* **2018**, *81*, 12–18. [CrossRef]

49. Wahlgren, N.M.; Dejmek, P.; Drakenberg, T. A 43Ca and 31P NMR study of the calcium and phosphate equilibria in heated milk solutions. *J. Dairy Res.* **1990**, *57*, 355–364. [CrossRef]

50. Pisanu, S.; Pagnozzi, D.; Pes, M.; Pirisi, A.; Roggio, T.; Uzzau, S.; Addis, M.F. Differences in the peptide profile of raw and pasteurised ovine milk cheese and implications for its bioactive potential. *Int. Dairy J.* **2015**, *42*, 26–33. [CrossRef]

51. Grappin, R.; Beuvier, E. Possible implications of milk pasteurization on the manufacture and sensory quality of ripened cheese. *Int. Dairy J.* **1997**, *7*, 751–761. [CrossRef]

52. Singh, H.; Waungana, A. Influence of heat treatment of milk on cheesemaking properties. *Int. Dairy J.* **2001**, *11*, 543–551. [CrossRef]

53. Slade, L.; Levine, H. Beyond water activity: Recent advances based on an alternative approach to the assessment of food quality and safety. *Crit. Rev. Food Sci. Nutr.* **1991**, *30*, 115–360. [CrossRef]

54. Brosnan, T.; Sun, D.-W. Inspection and grading of agricultural and food products by computer vision systems—A review. *Comput. Electron. Agric.* **2002**, *36*, 193–213. [CrossRef]

55. Narendra, V.G.; Hareesha, K.S. Quality inspection and grading of agricultural and food products by computer vision—A review. *Int. J. Comput. Appl.* **2010**, *2*. [CrossRef]

56. Wang, H.-H.; Sun, D.-W. Melting characteristics of cheese: Analysis of effect of cheese dimensions using computer vision techniques. *J. Food Eng.* **2002**, *52*, 279–284. [CrossRef]

57. Caccamo, M.; Melilli, C.; Barbano, D.M.; Portelli, G.; Marino, G.; Licitra, G. Measurement of gas holes and mechanical openness in cheese by image analysis. *J. Dairy Sci.* **2004**, *87*, 739–748. [CrossRef]

58. Ni, H.; Guansekaran, S. Image processing algorithm for cheese shred evaluation. *J. Food Eng.* **2004**, *61*, 37–45. [CrossRef]

59. Zhou, L.; Zhang, C.; Liu, F.; Qiu, Z.; He, Y. Application of deep learning in food: A review. *Compr. Rev. Food Sci. Food Saf.* **2019**, *18*, 1793–1811. [CrossRef]

60. Lundervold, A.S.; Lundervold, A. An overview of deep learning in medical imaging focusing on MRI. *Z. Med. Phys.* **2019**, *29*, 102–127. [CrossRef]

61. Weiss, K.; Khoshgoftaar, T.M.; Wang, D. A survey of transfer learning. *J. Big Data* **2016**, *3*, 9. [CrossRef]

Communication

GC-MS Metabolomics and Antifungal Characteristics of Autochthonous *Lactobacillus* Strains

Paola Scano [1,*], M. Barbara Pisano [2], Antonio Murgia [1], Sofia Cosentino [2] and Pierluigi Caboni [1]

[1] Department of Life and Environmental Sciences, University of Cagliari, SS 554 km 4.5, 09042 Monserrato, Italy; a-murgia@hotmail.it (A.M.); caboni@unica.it (P.C.)
[2] Department of Medical Sciences and Public Health, University of Cagliari, SS 554 km 4.5, 09042 Monserrato, Italy; barbara.pisano@unica.it (M.B.P.); scosenti@unica.it (S.C.)
* Correspondence: scano@unica.it; Tel.: +39-070-675-4391

Abstract: *Lactobacillus* strains with the potential of protecting fresh dairy products from spoilage were studied. Metabolism and antifungal activity of different *L. plantarum*, *L. brevis*, and *L. sakei* strains, isolated from Sardinian dairy and meat products, were assessed. The metabolite fingerprint of each strain was obtained by GC-MS and data submitted to multivariate statistical analysis. The discriminant analysis correctly classified samples to the *Lactobacillus* species and indicated that, with respect to the other species, *L. plantarum* had higher levels of organic acids, while *L. brevis* and *L. sakei* showed higher levels of sugars than *L. plantarum*. Partial Least Square (PLS) regression correlated the GC-MS metabolites to the antifungal activity ($p < 0.05$) of *Lactobacillus* strains and indicated that organic acids and oleamide are positively related with this ability. Some of the metabolites identified in this study have been reported to possess health promoting proprieties. These overall results suggest that the GC-MS-based metabolomic approach is a useful tool for the characterization of *Lactobacillus* strains as biopreservatives.

Keywords: *Lactobacillus*; antifungal activity; fresh cheese; biopreservation; GC-MS; metabolomics

Citation: Scano, P.; Pisano, M.B.; Murgia, A.; Cosentino, S.; Caboni, P. GC-MS Metabolomics and Antifungal Characteristics of Autochthonous *Lactobacillus* Strains. *Dairy* **2021**, 2, 326–335. https://doi.org/10.3390/dairy2030026

Academic Editor: Burim Ametaj

Received: 11 May 2021
Accepted: 18 June 2021
Published: 23 June 2021

Publisher's Note: MDPI stays neutral with regard to jurisdictional claims in published maps and institutional affiliations.

1. Introduction

In the Mediterranean area, to revive the ovine dairy industry, the development of new fresh dairy products is increasing. Taking into account that the physico-chemical characteristics of these products greatly favor the growth of pathogenic and spoilage microbes, many research activities have been devoted to finding solutions to protect fresh dairy products from deterioration and to prolong their shelf life while preserving the organoleptic and nutritional properties. Furthermore, in recent years, the increased request for natural foods has challenged the food producers and researchers to find natural alternatives to synthetic preservatives. Among natural food preservation techniques, bio-preservation has gained particular attention; this term refers to the use of microbial strains or metabolites able to inhibit the growth of spoilage and pathogenic microbes in foods and thus improving safety and extending their shelf-life [1]. Lactic acid bacteria (LAB), naturally present in many fermented and not fermented foods of animal and vegetable origin, are considered good candidates as bio-preservative, as they possess numerous functional properties related to the improvement of food safety and health benefits [2,3]. Among LAB, several strains belonging to the genus *Lactobacillus*, frequently present in milk and cheese as the prevalent nonstarter LAB, have shown distinct antimicrobial and antifungal activities [4]. Additionally, *Lactobacillus sakei* strains, characteristic of meat products, have shown antifungal activity [4]. The latter strains can be found in raw meat products stored under vacuum and refrigerated, as well as in fermented sausages, due to their metabolic properties and phenotypic features that are particularly well adapted to growth and survival under the conditions found during meat processing and storage [5,6].

The antifungal activity of LAB may be related to the synergistic action of several compounds, e.g., organic acids (acetic acid, lactic acid, propionic acid, and phenyllactic acid), hydrogen peroxide, cyclic dipeptides, proteinaceous compounds (bacteriocins), and fatty acids [2,7], and it is recognized that the capacity to synthetize these compounds is a strain-linked trait [4]. In our previous work, we found that selected *Lactobacillus* strains of food origin could control mold contamination without altering the sensory characteristics in miniature Caciotta ovine cheese, produced in a laboratory scale [4]; one limitation of this study is that the compounds responsible for the antifungal activity were not studied. Metabolomics is one of the most valuable techniques for the study of metabolite profiles in food matrices [8], and gas chromatography coupled with mass spectrometry (GC-MS) is a well-suited analytical platform for the study of microbial metabolites [9,10]. By the combined approach of GC-MS and multivariate statistical data analysis it was possible to find links between metabolites of dairy products and their microbial profiles [11]. The aim of this work was to test the suitability of this approach for the identification of the metabolites involved in the bioprotective characteristics of autochthonous *Lactobacillus* strains isolated from Sardinian dairy and meat products and to seek correlations between metabolites (or group of metabolites) and antifungal activity. To this goal, we studied the polar and semi-polar low molecular weight metabolites present in the cultured broth of different *Lactobacillus* strains. Antifungal activity of these strains against mold species, commonly occurring in cheese spoilage, was also assessed and results correlated to the GC-MS data by multivariate regression analysis.

2. Materials and Methods

2.1. Microorganisms and Cultivation Conditions

A total of 9 autochthonous *Lactobacillus* strains (3 *Lactobacillus plantarum*, 2 *Lactobacillus brevis*, and 4 *Lactobacillus sakei*), deposited in the MBDS culture collection (www.mbds.it, accessed on 25 May 2021) were studied. They were isolated from ovine raw milk and artisanal cheeses (*L. plantarum* and *L. brevis*), and sausages (*L. sakei*) manufactured in Sardinia (Table 1), and they were classified on the basis of phenotypic characteristics and molecular methods through polymerase chain reaction amplification. *Lactobacillus* strains were stored at −20 °C in DeMan Rogosa Sharpe (MRS) broth (Microbiol, Cagliari, Italy) with 15% (*v*/*v*) glycerol and routinely cultivated on MRS agar plates for 48 h at 30 °C in microaerophylia. Each strain was subcultered twice in MRS broth prior to experimental use. Carbohydrate fermentation profiles of the *Lactobacillus* strains was assessed and reported in Table S1.

Table 1. List of the *Lactobacillus* strains and food origin.

Species	Strains	Origin	Molecular Identification	MBDS #
Lactobacillus plantarum	4/16898	Raw sheep's milk	Species- specific PCR [12]	UNICAB27
Lactobacillus plantarum	1/14537	Raw sheep's milk	Species- specific PCR [12]	UNICAB32
Lactobacillus plantarum	C1/15	Sheep's cheese	Species- specific PCR [12]	UNICAB212
Lactobacillus brevis	DSM 32516	Sheep's cheese	16S rRNA gene sequencing Universal primers F357-R518 [13]	UNICAB24
Lactobacillus brevis	M8/1	Sheep's cheese	16S rRNA gene sequencing Universal primers F357-R518 [13]	UNICAB456
Lactobacillus sakei	S3	Artisanal sausage	16S rRNA gene sequencing Universal primers Y1-Y2 [14]	UNICAB457
Lactobacillus sakei	S5	Artisanal sausage	16S rRNA gene sequencing Universal primers Y1-Y2 [14]	UNICAB458
Lactobacillus sakei	S4	Artisanal sausage	16S rRNA gene sequencing Universal primers F357-R518 [13]	UNICAB459
Lactobacillus sakei	S3/1	Artisanal sausage	16S rRNA gene sequencing Universal primers Y1-Y2 [14]	UNICAB460

2.2. Extraction of Extracellular Samples

Cell-free supernatants were obtained from approximately 15 mL of overnight cultures of the *Lactobacillus* strains in MRS broth at 30 °C. Then, the cells were separated by centrifugation at 7000× g, for 20 min at 4 °C and supernatant was taken and filtered through a cellulose acetate membrane filter (0.2 μm pore size) to eliminate residual bacterial cells. Cell-free supernatant was distributed in 1 mL aliquots, then dH$_2$O (10 mL) and internal standard (0.2 μmol of 10 mmol solution of 2,3,3,3-d4 D,L-alanine) were added to each sample. Three replicate samples were prepared for each *Lactobacillus* strain cell-free supernatant. Un-inoculated MRS broth samples were collected as control. An aliquot of 100 μL of each sample was stored at −80 °C.

2.3. In Vitro Antifungal Activity of Lactobacillus

Antifungal activity of *Lactobacillus* strains against *Alternaria alternata* (UNICAM2), *Cladosporium herbarum* (UNICAM10), *Paecilomyces variotii* (UNICAM17), and *Penicillium chrysogenum* (ATCC 9179) indicator strains was assessed in vitro using the agar plate method previously described [4]. The inhibitory activity against the fungal strains *Aspergillus flavus* (ATCC 46283), *Fusarium oxysporum* (UNICAM12), and *Mucor recurvus* (UNICAM15), which develop faster than lactobacilli, was assessed using the dual-culture overlay assay [4]. For all assays, the antifungal activity of each strain was determined by measuring the width of the halo around the bacterial streaks, according to the following 4 steps in a semiquantitative scale: (1) ≥8 mm, (2) 5–7 mm, (3) 3–4 mm, and (4) <3 mm.

2.4. Sample Preparation for GC-MS Analysis

Samples were thawed on ice and subsequently 375 μL of a methanol and chloroform mixture (2/1, v/v) was added. After 1 h, 380 μL of chloroform and 90 μL of aqueous KCl 0.2 M were added. Samples were vortexed for 10 s and centrifuged at 15,294× g for 10 min at 4 °C. A volume of 200 μL of the aqueous fraction was taken and transferred into 1.5 mL sterile glass vials with 10 μL of a 80 mg/L of 2,2,3,3-d4-succinic acid solution. The aqueous fraction was then dried by a nitrogen stream, and derivatized with 40 μL of MSTFA (N-Trimethylsilyl-N-methyl trifluoroacetamide). Samples were kept for 30 min at 70 °C, before adding 600 μL of hexane and then vortexed.

2.5. GC-MS Analysis

Samples were analyzed with a Hewlett Packard 6850 Gas Chromatograph, 5973 mass selective detector, and 7683B series injector (Agilent Technologies, Palo Alto, CA, USA) with helium as carrier gas at a flow of 1.0 mL/min. One microliter of each sample was injected with 1 min of split flow delay and resolved on a 30 m × 0.25 mm × 0.25 μm DB-5MS column (Agilent Technologies, Palo Alto, CA, USA). Inlet, interface, and ion source temperatures were 250, 250 and 230 °C, respectively. Oven starting and final temperatures were set at 50 and 230 °C, respectively, with a rate of 5 °C/min for 36 min and then for 2 min at a constant temperature. Electron impact mass spectra were recorded from m/z 50 to 550 at 70 eV. Metabolite annotation was achieved by mass spectra comparison with analytical standards, in house library and the NIST14 database (National Institute of Standards and Technology, Gaithersburg, MD, USA).

2.6. Multivariate Statistical Data Analysis

As input for multivariate statistical data analysis (MVA), a 27 × 59 matrix, composed of samples × the intensities of the most abundant fragment ion for each metabolite, was constructed. MVA were performed using the software SIMCA-P+ (version 14.1, Umetrics, Umeå, Sweden). GC-MS variables were mean centered and scaled to unit variance. The principal component analysis (PCA) was carried out to study the sample distributions and presence of outliers. For classification of samples to the three classes of *Lactobacillus*, and to find discriminant metabolites, the supervised partial least squares-discriminant analysis (PLS-DA) was used. Importance of the variables in PLS-DA was assessed by considering the variable importance in the projection (VIP) values. Results of PLS-DA are shown as scatter plot of scores and loadings in the first two principal components, and as a list of metabolites having VIP > 1 attributed to each class of samples on the basis of their coefficients. In the Coefficients Overview Plot the X-variable coefficients, together with their standard errors in cross-validation, for the Y-class are shown. The supervised partial least squares (PLS) regression was applied for the study of linear relationships between antifungal activity and metabolite profiles. Antifungal activity scores were organized in the Y matrix. Results are shown as PLS loading weights plot in first and second components that displays the relation between the X-variables (GC-MS data) and the Y-variables (antifungal activity). Metabolites positively correlated with antifungal high activity lie next to the Y-variable, metabolites negatively correlated lie in the opposite side of the plot. The quality of the PLS and PLS-DA models and the proper number of principal components were assessed on the basis of the cumulative parameters R^2Y and Q^2Y (prediction power calculated in cross-validation), and of their difference value, with a <0.50 threshold.

3. Results

3.1. GC-MS Metabolite Profiles of Lactobacillus Species

The hydrophilic extracts of samples were analyzed by GC-MS and 59 low molecular weight metabolites were annotated and reported in Table S2. Short-chain carboxylic acids (such as lactic acid, citric acid, succinic acid, and phenyllactic acid), amino acids (valine, leucine, isoleucine, threonine, and pyroglutamic acid), mono- and disaccharides, and polyols were annotated together with fatty acids and analogues, such as oleamide. Analysis of MRS broth chromatograms indicates that it was mainly composed by mono- and disaccharides, citric acid, phosphate, and amino acids essential for growth (results not shown).

To assess differences among the three species of *Lactobacillus*, a PLS-DA was carried out. The discriminant analysis, with a high degree of confidence, well classified samples (score plot in Figure 1a), indicating that the 3 *Lactobacillus* species showed overall different metabolite profiles. By analysis of the loading plot (Figure 1b) and metabolite VIP scores reported in Table 2 (coefficients of metabolites are shown in Figure S1), the fermentation medium of *L plantarum* was found to contain more lactic acid, 3-hydroxy butyric acid (BHBA), 2-hydroxy isovaleric acid (AHVA), 2-hydroxy isocaproic acid (AHCA), succinic acid, 3-phenyllactic acid, and malic acid, together with lower levels of mono- and disaccharides than *L. brevis* and *L. sakei*. In contrast, *L. sakei* showed more sugars. When compared to the other two *Lactobacillus* species, *L. brevis* had more monosaccharides, AHVA, and 4-gamma-aminobutyric acid (GABA) and *L. sakei* showed higher levels of disaccharides. *L. plantarum* and *L. sakei* showed higher levels of pyroglutamic acid. The higher level of saccharides in *L. sakei* and *L. brevis* than in *L. plantarum* was confirmed by the measured carbohydrates fermentation profiles reported in Table S1, where the *L. plantarum* strains, compared to the other 2 species, showed the better fermentation activity.

Broth sugar consumption from *Lactobacillus* strains and production of organic acids are confirmed by Pearson correlation values between GC-MS data depicted as a heat map in Figure S2. In this map, it is clearly visible that organic acids were strongly positively correlated with themselves (r > 0.75) and negatively correlated (r < 0.75) with saccharides (the linear strong negative relationship r = −0.97 between glucose and lactic acid is clearly visible in the Figure S3).

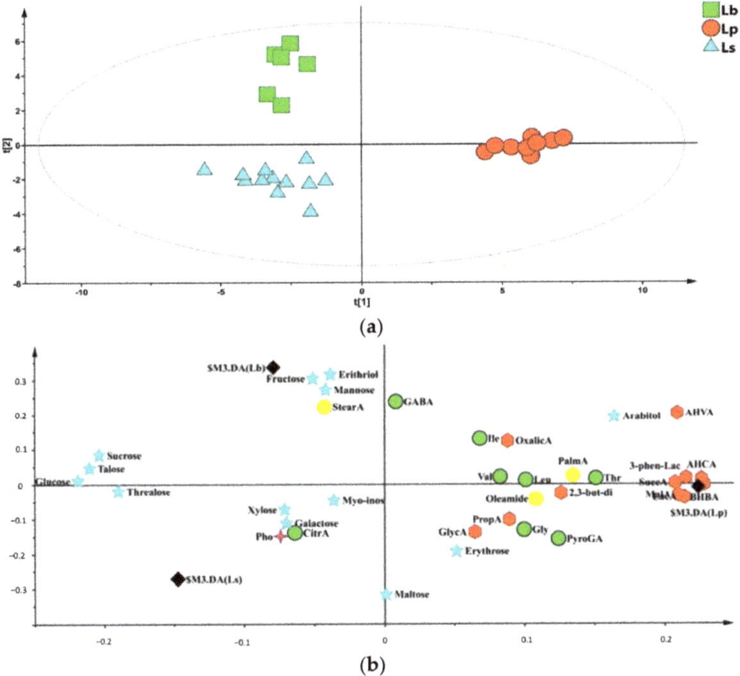

Figure 1. PLS-DA of GC-MS data, $R^2Y = 0.93$ and $Q^2Y = 0.88$, over 2 validated components. (**a**) score plot (*Lb*, *Lp*, and *Ls* = *L. brevis*, *L. plantaris*, and *L. sakei*, respectively); (**b**) loading plot, red hexagons = organic acids; green circles = amino acids; light blue stars = saccharides and polyols; yellow circles = fatty acids and analogues; black diamond = loadings of *Lactobacillus* classes. Unknown metabolites are not displayed. Metabolites are abbreviated as in Table S2.

Table 2. PLS-DA VIP [a] scores of discriminant GC-MS metabolites.

L. plantarum [b]			*L. brevis*			*L. sakei*		
Metabolite	Class [c]	VIP	Metabolite	Class	VIP	Metabolite	Class	VIP
3-hydroxy butyric acid (BHBA)	OA	1.25	Fructose	S	1.66	Maltose	S	1.65
2-hydroxy isovaleric acid (AHVA)	OA	1.51	Erithritol	S	1.64	Sucrose	S	1.23
Arabitol	S	1.32	2-hydroxy isovaleric acid (AHVA)	OA	1.51	Glucose	S	1.20
2-hydroxy isocaproic acid (AHCA)	OA	1.24	Mannose	S	1.45	Talose	S	1.20
Lactic Acid	OA	1.20	Arabitol	S	1.32	Pyroglutamic acid	AA	1.09
3-phenyllactic acid	OA	1.18	4-aminobutyric acid (GABA)	AA	1.24	Erythrose	S	1.04
Malic acid	OA	1.17	Sucrose	S	1.23	Threalose	S	1.03
Succinic acid	OA	1.13	Stearic acid	FA	1.21			
Pyroglutamic acid	AA	1.09	Talose	S	1.20			
Erythrose	S	1.04	Threalose	S	1.03			

[a] Variable importance in the projection, $R^2Y = 0.93$ and $Q^2Y = 0.88$ over 2 validated components; [b] *Lactobacillus* species with higher level of the metabolite, as indicated by the coefficients; [c] OA = organic acids, AA = amino acids and analogues, S = saccharides and polyols, FA = fatty acids and analogues.

3.2. Metabolite Profiles of Lactobacillus Strains

A visual analysis of chromatograms suggests that all the strains consumed sugars and produced lactic acid, though to a different extent, and synthetized oleamide. *L. plantarum* 1/14537 and *L. brevis* DSM 32516 consumed almost all the citric acid of broth origin (data not shown). To study the different production/consumption of metabolites by the nine *Lactobacillus* strains, metabolites indicated by the PLS-DA as the most discriminant between the three species, together with oleamide, were individually measured and results reported as column plot with means and standard deviations (Figure 2). Phenyllactic acid, AHCA, BHBA, AHVA, and malic acid were produced at a greater extent by *L. plantarum*. Pyroglutamic acid was mainly produced by all the *L. plantarum* strains and *L. sakei* S3/1. *L. brevis* M8/1 produced the greatest, by far, quantity of GABA, confirming that different *Lactobacillus* strains within the same species can have their own individual metabolism.

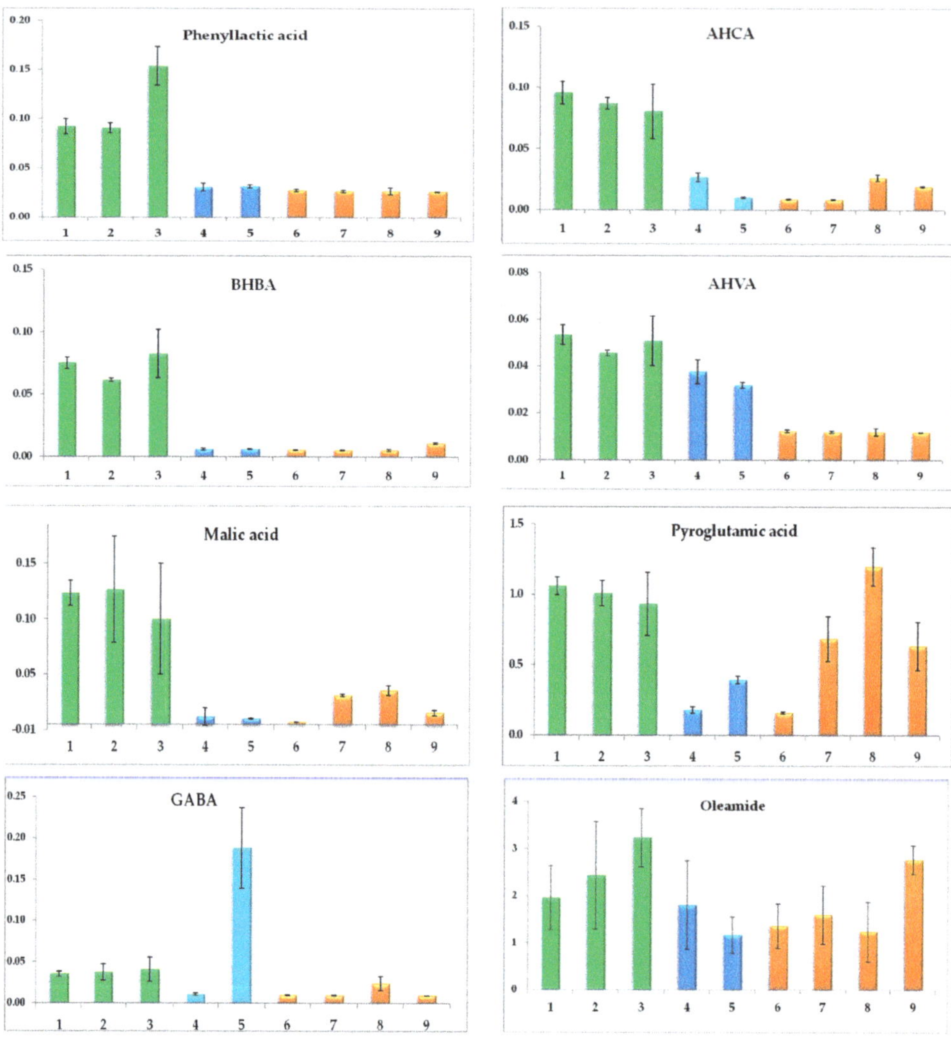

Figure 2. Relative abundance of metabolites in each strain: green *L. plantarum* (1 = 1/14537, 2 = 4/16898, 3 = C1/15); light blue *L. brevis* (4 = DSM 32516, 5 = M8/1 S4); yellow *L. sakei* (6 = S4, 7 = S3, 8 = S3/1, 9 = S5). Y-axis values are in A.U. and refer to the intensity of a selected *m/z* ion fragment. Abbreviation of metabolites as in Table S2.

3.3. Antifungal Activity and Correlations with Metabolite Profiles

Antifungal activity of *Lactobacillus* strains was also studied. Seven mold species were chosen for their widespread presence in cheese spoilage and for their ability to produce mycotoxins [4]. As shown in Figure 3, the strains exhibited a wide range of antifungal activity, dependent on both fungal species and *Lactobacillus* strain. All *L. plantarum* strains and *L. brevis* DSM 32516 showed the strongest activity with inhibition zones greater than 8 mm against all the fungal strains. The *L. sakei* strains were the less active, with S4 strain showing inhibition zones less than 3 mm against *M. recurvus*, *P. variotii*, and *P. chrysogenum*. *A. alternata* and *C. herbarum* growth was strongly affected (inhibition zone higher than 8 mm) by all the *Lactobacillus* strains except for *L. sakei* S5 which had an inhibition zone lower than 8 mm.

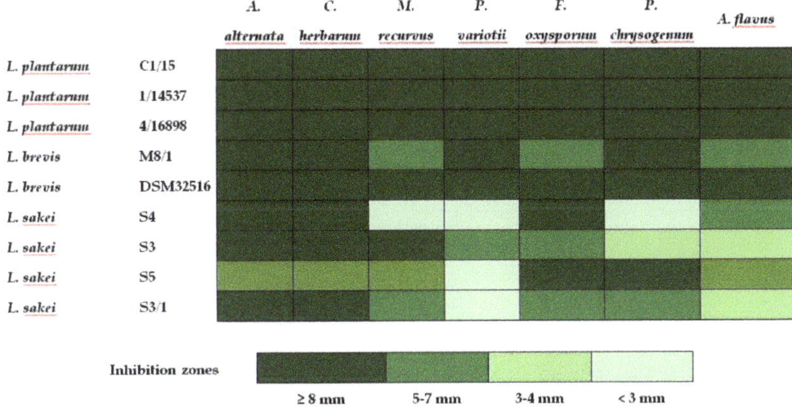

Figure 3. Antifungal activity of *Lactobacillus* strains: inhibition zones against *A. alternata*, *C. herbarum*, *M. recurvus*, *P. variotii*, *F. oxysporum*, *A. flavus* ATCC 46283, *P. chrysogenum* ATCC 9179.

These results are in agreement with a previous study [4] where single *L. plantarum* C1/15 strain or multiple *Lactobacillus* strains (*L. plantarum* 4/16898, 1/14537 and *L. brevis* DSM 32516) were able to significantly inhibit the growth of *P. chrysogenum* and *A. flavus* when used as adjunct cultures in the production of miniature Caciotta cheese. With respect to the single *L. plantarum* strain (C1/15), the multiple *Lactobacillus* strain combination resulted more active against the fungal strains tested confirming the synergistic action of different antimicrobial compounds highlighted in the current study [4]. To correlate the metabolite profile of strains to the antifungal activity a PLS regression of GC-MS data was carried out considering the extent of the antifungal activity as Y variable. The resulting loading plot in the first two principal components with superimposed scores of antifungal activity is shown in Figure 4.

In this plot, metabolites that lie next to the antifungal activity have positive relationship with this latter, metabolites in the opposite side of the plot, passing from the origin are inversely related with the activity. On the basis of these rules, we can note that organic acids and arabitol were directly involved in the activity against *P. chrysogenum*, *P. variotii*, and *M. recurvus*. Oleamide was correlated with *A. flavus* and *F. oxysporum*.

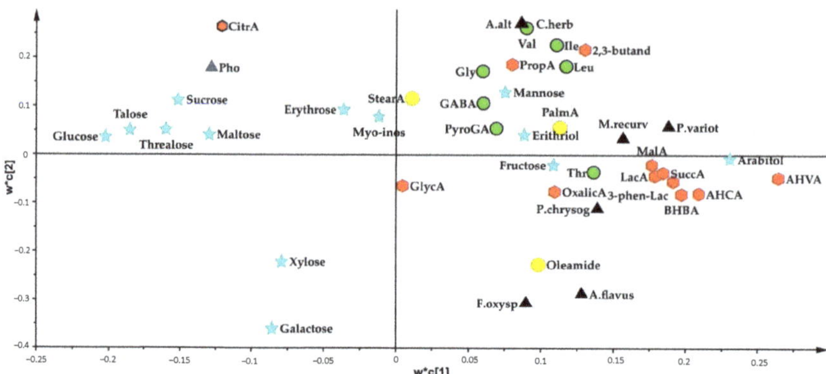

Figure 4. PLS variable loadings plot of GC-MS data (X-variables) superimposed with antifungal activity (Y-variables). $R^2Y = 0.89$ and $Q^2Y = 0.65$ over 5 validated components. Black triangles = Y coefficients for antifungal activity; red hexagons = organic acids; green circles = amino acids; light blue stars = saccharides and polyols; yellow circles = fatty acids and analogues; black diamond = loadings of *Lactobacillus* classes. Unknown metabolites are not displayed. Metabolites are abbreviated as in Table S2.

4. Discussion

The inherent ability of *Lactobacillus* to produce compounds antagonistic to food microbial contaminants as a part of their natural defense mechanism can provide the dairy food industry with natural and healthier alternatives to chemical preservatives [15], especially for fresh products. The antifungal activity of *Lactobacillus* has been related to the synergistic effect of several compounds, and part of these compounds shows nutraceutical properties.

Compared to the other species here studied, *L. plantarum* strains showed more organic acids, among which 3-phenyllactic acid, which shows a broad spectrum of antimicrobial properties, is effective against bacteria and fungi including yeast [16]. In sourdough bread started with *L. plantarum* 21B, the onset of growth of *Aspergillus niger* was delayed 7 days, with respect to bread started with a *L. brevis* strain not producing phenyllactic acid [17]. However, it has been reported that phenyllactic acid is active against fungal species only at mg mL^{-1} concentrations, suggesting an overall antifungal effect in association with other factors produced by LAB [17]. 3-Phenyllactic acid is a by-product of phenylalanine metabolism in *Lactobacillus*, where phenylalanine is transaminated to phenylpyruvic acid and further reduced to 3-phenyllactic acid by 2-hydroxy dehydrogenase [15]. GABA is another product of amino acid metabolism, biosynthesized by microorganisms through decarboxylation of glutamate by glutamate decarboxylase [18]. GABA is an important bioactive compound and its production by various microorganisms has been actively explored [18]. Higher levels of GABA were detected solely in the *L. brevis* M8/1 strain. Besides lactic acid, a number of 2-hydroxy organic acids, that show antimicrobial properties [10], are generally produced by *Lactobacillus*. Organic acids levels were found correlated with the antifungal activity against *P. chrysogenum*, *P. variotii*, and *M. recurves*. *L. plantarum* was found to produce hydroxy organic acids, some of which (AHVA and AHCA) have shown antifungal proprieties [10]. In particular, AHCA has been studied for its potential in inhibiting cell growth and biofilm formation of the pathogenetic *Aspergillus* and *Candida* species in topical use [19,20]. One of the most important virulence factors of *C. albicans* is the ability to form biofilms that protect this yeast against endogenous and exogenous inhibitory substances, in fact almost 65% of microbial infections in humans are biofilm-related [19]. The effects of oral administration of lactobacilli against biofilm-associated infections have been studied [21]. All the strains synthetized oleamide, not present in the broth, and levels of this compounds have been correlated with the antifungal activity against *A. flavus* and *F. oxysporum*. Oleamide in an amide of oleic acid present in small amount in animal brains and can be produced by microorganisms [22]. In dairy products' fermentation processes, oleamide is synthesized from oleic acid, which is an abundant component in these food products, owing to lipase enzymatic amidation [23]. Epidemiological and clinical data have shown that fermented dairy products possess preventive effects against dementia, including Alzheimer's disease, and those effects have been linked to oleamide [23]. This compound has been identified as the agent responsible for reducing microglial inflammatory responses and neurotoxicity [23]. Oleamide is also an endogenous bioactive signaling molecule that acts in various cell types and could elicit different biological effects [23]. Among the wide range of functions, the most acknowledged one is its sleep-inducing effect [24]. It has been found that *L. plantarum* 1/14537 and *L. brevis* DSM 32516 consumed almost all the citric acid of broth origin. Citric acid fermentation by LAB could lead to acetoin and diacetyl production which are aromatic compounds. However, obligately heterofermentative lactobacilli such as *L. brevis* when present at high levels and facultatively heterofermentative lactobacilli such as *L. plantarum*, under certain conditions, by co-metabolization of citrate and different sugars may produce great amount of CO_2, giving rise to holes, splits and blowing defects of aged cheeses [25]. *L. brevis* strains are also characterized by other biochemical properties such as production of CO_2 from glucose and ammonia from arginine, as shown in Table S1, that are generally used as tool for lactobacilli classification [26]. Moreover, *L. brevis* could contribute to the degradation of arginine in cheese that can lead to ornithine accumulation by the arginine-urease or the arginine deaminase pathways [27]. On the other hand, ornithine and other free amino acids could be decarboxylated to putrescine and other biogenic amines that in high concentration could be responsible for toxic effect on human health.

In conclusion, results of this study provide reference data for interpreting the differences in the metabolite profiles of *Lactobacillus* strains isolated from different fermented food products and their correlation with antifungal activity in vitro. These autochthonous strains could be considered potential good candidates to be used in cheese manufacturing as bioprotective cultures and for in situ production of postbiotic substances. However, further studies in cheese model systems are warranted to evaluate in situ positive metabolic activities such as organic acids production, GABA and oleamide formation, as well as negative activities including biogenic amines production and the potential to cause blowing defects in the initial and later stages of ripening.

Supplementary Materials: The following are available online at https://www.mdpi.com/article/10.3390/dairy2030026/s1, Figure S1: PLS-DA coefficient overview plot; Figure S2: Heat map of Pearson correlation values of GC-MS data matrix; Figure S3: Correlation plot of lactic acid (X-axis) vs. glucose (Y-axis) levels. Table S1: Carbohydrates fermentation profiles of Lactobacillus strains; Table S2: List of GC-MS metabolites.

Author Contributions: Conceptualization, P.C., M.B.P., S.C., and P.S.; Methodology, P.S., A.M., and M.B.P.; Software, P.S.; Validation, P.S. and M.B.P.; Formal Analysis, P.S., A.M., and M.B.P.; Writing—Original Draft Preparation, P.S., P.C. and M.B.P.; Writing—Review and Editing, P.S., P.C., S.C. and M.B.P. All authors have read and agreed to the published version of the manuscript.

Funding: This research received no external funding.

Informed Consent Statement: Not applicable.

Data Availability Statement: Not report data available.

Conflicts of Interest: The authors declare no conflict of interest.

References

1. Stiles, M.E. Biopreservation by lactic acid bacteria. *Antonie Van Leeuwenhoek* **1996**, *70*, 331–345. [CrossRef]
2. Crowley, S.; Mahony, J.; van Sinderen, D. Current perspectives on antifungal lactic acid bacteria as natural bio-preservatives. *Trends Food Sci. Technol.* **2013**, *33*, 93–109. [CrossRef]
3. Schnürer, J.; Magnusson, J. Antifungal lactic acid bacteria as biopreservatives. *Trends Food Sci. Technol.* **2005**, *16*, 70–78. [CrossRef]
4. Cosentino, S.; Viale, S.; Deplano, M.; Fadda, M.E.; Pisano, M.B. Application of autochthonous Lactobacillus strains as biopreservatives to control fungal spoilage in Caciotta cheese. *BioMed. Res. Int.* **2018**, *2018*. [CrossRef] [PubMed]

5.   Champomier-Vergès, M.C.; Chaillou, S.; Cornet, M.; Zagorec, M. Lactobacillus sakei: Recent developments and future prospects. *Res. Microbiol.* **2001**, *152*, 839–848. [CrossRef]
6.   Zagorec, M.; Champomier-Vergès, M.C. Lactobacillus sakei: A starter for sausage fermentation, a protective culture for meat products. *Microorganisms* **2017**, *5*, 56. [CrossRef]
7.   Dalié, D.K.D.; Deschamps, A.M.; Richard-Forget, F. Lactic acid bacteria–Potential for control of mould growth and mycotoxins: A review. *Food Control* **2010**, *21*, 370–380. [CrossRef]
8.   Cevallos-Cevallos, J.M.; Reyes-De-Corcuera, J.I.; Etxeberria, E.; Danyluk, M.D.; Rodrick, G.E. Metabolomic analysis in food science: A review. *Trends Food Sci. Technol.* **2009**, *20*, 557–566. [CrossRef]
9.   Chaudhary, A.; Verma, K.; Saharan, B.S. A GC-MS Based Metabolic Profiling of Probiotic Lactic Acid Bacteria Isolated from Traditional Food Products. *J. Pure Appl. Microbiol.* **2020**, *14*, 657–672. [CrossRef]
10.   Siedler, S.; Balti, R.; Neves, A.R. Bioprotective mechanisms of lactic acid bacteria against fungal spoilage of food. *Curr. Opin. Biotechnol.* **2019**, *56*, 138–146. [CrossRef] [PubMed]
11.   Pisano, M.B.; Scano, P.; Murgia, A.; Cosentino, S.; Caboni, P. Metabolomics and microbiological profile of Italian mozzarella cheese produced with buffalo and cow milk. *Food Chem.* **2016**, *192*, 618–624. [CrossRef]
12.   Quere, F.; Deschamps AUrdaci, M.C. DNA probe and PCR-specific reaction for *Lactobacillus plantarum*. *J. Appl. Microbiol.* **1997**, *82*, 783–790. [CrossRef] [PubMed]
13.   Muyzer, G.; de Waal, E.C.; Uitterlinden, A.G. Profiling of complex microbial populations by denaturing gradient gel electrophoresis analysis of polymerase chain reaction-amplified genes coding for 16S rRNA. *Appl. Environ. Microbiol.* **1993**, *59*, 695–700. [CrossRef]
14.   Young, J.P.W.; Downer, H.L.; Eardly, B.D. Phylogeny of the phototrophic Rhizobium strain BTAil by polymerase chain reaction-based sequencing of a 16S rRNA gene segment. *J. Bacteriol.* **1991**, *173*, 2271–2277. [CrossRef] [PubMed]
15.   Naz, S.; Cretenet, M.; Vernoux, J.P. Current Knowledge on Antimicrobial Metabolites Produced from Aromatic Amino Acid Metabolism in Fermented Products. In *Microbial Pathogens and Strategies for Combating Them: Science, Technology and Education*; Méndez-Vilas, A., Ed.; Formatex Research Center: Badajoz, Spain, 2013; pp. 337–346.
16.   Chaudhari, S.S.; Gokhale, D.V. Phenyllactic acid: A potential antimicrobial compound in lactic acid bacteria. *J. Bacteriol. Mycol. Open Access* **2016**, *2*, 00037. [CrossRef]
17.   Lavermicocca, P.; Valerio, F.; Evidente, A.; Lazzaroni, S.; Corsetti, A.; Gobbetti, M. Purification and characterization of novel antifungal compounds from the sourdough Lactobacillus plantarum strain 21B. *Appl. Environ. Microbiol.* **2000**, *66*, 4084–4090. [CrossRef]
18.   Tajabadi, N.; Baradaran, A.; Ebrahimpour, A.; Rahim, R.A.; Bakar, F.A.; Manap, M.Y.A.; Mohammed, A.S.; Saari, N. Overexpression and optimization of glutamate decarboxylase in Lactobacillus plantarum Taj-Apis 362 for high gamma-aminobutyric acid production. *Microbial. Biotechnol.* **2015**, *8*, 623–632. [CrossRef]
19.   Nieminen, M.T.; Novak-Frazer, L.; Rautemaa, V.; Rajendran, R.; Sorsa, T.; Ramage, G. A Novel Antifungal Is Active against Candida albicans Biofilms and Inhibits Mutagenic Acetaldehyde Production In Vitro. *PLoS ONE* **2014**, *9*, e97864. [CrossRef]
20.   Sakko, M.; Moore, C.; Novak-Frazer, L.; Rautemaa, V.; Sorsa, T.; Hietala, P.; Järvinen, A.; Bowyer, P.; Tjäderhane, L.; Rautemaa, R. 2-hydroxyisocaproic acid is fungicidal for Candida and Aspergillus species. *Mycoses* **2014**, *57*, 214–221. [CrossRef]
21.   Vuotto, C.; Longo, F.; Donelli, G. Probiotics to counteract biofilm-associated infections: Promising and conflicting data. *Int. J. Oral Sci.* **2014**, *6*, 189–194. [CrossRef] [PubMed]
22.   Kwon, J.H.; Hwang, S.E.; Han, J.T.; Kim, C.J.; Rho, J.R.; Shin, J.E. Production of oleamide, a functional lipid, by Streptomyces sp. KK90378. *J. Microbiol. Biotechnol.* **2001**, *11*, 1018–1023.
23.   Ano, Y.; Nakayama, H. Preventive effects of dairy products on dementia and the underlying mechanisms. *Int. J. Mol. Sci.* **2018**, *19*, 1927. [CrossRef] [PubMed]
24.   Cravatt, B.F.; Prospero-Garcia, O.; Siuzdak, G.; Gilula, N.B.; Henriksen, S.J.; Boger, D.L.; Lerner, R.A. Chemical characterization of a family of brain lipids that induce sleep. *Science* **1995**, *268*, 1506–1509. [CrossRef]
25.   Blaya, J.; Barzideh, Z.; LaPointe, G. Symposium review: Interaction of starter cultures and nonstarter lactic acid bacteria in the cheese environment. *J. Dairy Sci.* **2018**, *101*, 3611–3629. [CrossRef] [PubMed]
26.   Axelsson, L. Lactic Acid Bacteria: Classification and Physiology. In *Lactic acid Bacteria, Microbiological and Functional Aspects*; Salminem, S., von Wright, A., Ouwehand, A., Eds.; Marcel Dekker: New York, NY, USA, 2004; Volume 139, pp. 1–66.
27.   Diana, M.; Rafecas, M.; Arco, C.; Quñez, J. Free amino acid profile of Spanish artisanal cheese: Importance of gamma-aminobutyric acid (GABA) and ornithine content. *J. Compos. Anal.* **2014**, *35*, 94–100. [CrossRef]

MDPI
St. Alban-Anlage 66
4052 Basel
Switzerland
Tel. +41 61 683 77 34
Fax +41 61 302 89 18
www.mdpi.com

Dairy Editorial Office
E-mail: dairy@mdpi.com
www.mdpi.com/journal/dairy

www.ingramcontent.com/pod-product-compliance
Lightning Source LLC
LaVergne TN
LVHW070613170726
843515LV00004B/67